Cochlear Mechanisms and Otoacoustic Emissions

Advances in Audiology

Vol. 7

Series Editor
M. Hoke, Münster

KARGER

Basel · München · Paris · London · New York · New Delhi · Bangkok · Singapore · Tokyo · Sydney

2nd International Symposium on Cochlear Mechanics and Otoacoustic Emissions,
Rome, March 9–11, 1989

Cochlear Mechanisms and Otoacoustic Emissions

Volume Editors
F. Grandori, Milan
G. Cianfrone, Rome
D.T. Kemp, London

99 figures and 8 tables, 1990

KARGER

Basel · München · Paris · London · NewYork · New Delhi · Bangkok · Singapore · Tokyo · Sydney

Advances in Audiology

© Copyright 1990 by S. Karger AG, P.O. Box, CH–4009 Basel (Switzerland)
Printed in Switzerland by Thür AG Offsetdruck, Pratteln
ISBN 3–8055–5082–0

Contents

Preface

Few areas of audiology have advanced as rapidly as cochlear physiology and biophysics has over the past decade. The advance began with the shock realization that existing knowledge and accepted concepts could not explain the response of the cochlea to sound and in particular otoacoustic emissions. Our very understanding of both the physical basis of hearing and the nature of hearing impairment was challenged.

The challenge was met from many directions by researchers holding views which were very much in the minority before 1980 and this work resulted in dramatic technical progress and discoveries. Hair cell motility was observed: nonlinear, sharp tuning in the intact mammalian cochlea was confirmed as the norm at low levels rather than the exception, otoacoustic emissions were widely studied and used both as a research and a clinical tool, the action of the cochlear efferent system was at last observed in the human ear. Until 1980, the cochlea was being modeled as a linear, passive, broadly tuned mechanical structure, as envisaged by von Békésy forty years ago. Otoacoustic emissions were a curious epiphenomenon of the transduction process. Sensory hearing impairment had nothing to do with cochlear mechanics. But, by the mid 1980s, the term 'active mechanism' had come to represent the process by which the whole mechanical response of the cochlea to sound is actually under the control of cochlear physiology, thus linking together hair cell motility, cochlear nonlinearity, frequency selectivity, otoacoustic emissions and sensory hearing loss. Much time was devoted to the discussion of these developments, and it seemed for a time that almost any new cochlear observation had to be related to the 'active process' if it were to gain respect!

These were major changes in thinking, but what have we actually gained from this decade of upheaval? Has audiology actually progressed? Fewer people would claim to understand the cochlear today than 10 years ago. For example, theoreticians are only beginning to explore the implications of the 'nonlinear and active process' for their calculations and have not yet been able to synthesize the new experimental data and concepts into a workable model. Is our concept right? This is a note of caution too from physiologists and biophysicists that hair cell motility, as presently observed, has not yet been proved to be the basis of the mechanical amplifier although this is now widely accepted as essential to cochlear function at low levels. But this hesitation to wholy embrace what seems so obviously right at this time is another sign of the very healthy state of audiological research and a good indicator of yet more progress to come. The cochlea has surprised us so many times in the past, and will do so again.

But there have been tremendous gains already. Without doubt auditory research is now well focused on the crucial issues related both to basic cochlear mechanisms and hearing disorder. Already new otoacoustic tools for screening and investigation have emerged and are in use.

This present volume, which basically arose out a Symposium on cochlear mechanics and the applications of otoacoustic emissions in Rome in March 1989, mirrors the 'state of the art' at the end of this most exciting decade, in a way that no other compilation has. It links the theoretical and the experimental with the practical business of audiology. From complementary positions in the field, some papers explain the problems facing those who would wish to theoretically understand the mechanical behavior of the cochlea, and leading cell biologists and biophysicists give an account of their latest experimental observations of hair motility, and of organ of Corti and basilar membrane motion. The curious properties of otoacoustic emissions are explored by several workers, presenting some new observations on transient evoked emissions, spontaneous and distortion product emissions. But the study of otoacoustic emissions is no longer just an exercise in auditory phenomenology. Nearly half the papers in this book directly address actual or potential clinical applications of otoacoustic emissions. Other aspects of cochlear pathology are not neglected: important new observations on tinnitus and frequency selectivity are reported as well.

Cochlear processing may still hold many of its secrets, but we are at last able to include these vital aspects of the hearing mechanisms in the audiological test battery an effort to understand auditory functions and alleviate hearing impairment.

F. Grandori, G. Cianfrone, D.T. Kemp

Grandori F, Cianfrone G, Kemp DT (eds): Cochlear Mechanisms and
Otoacoustic Emissions. Adv Audiol. Basel, Karger, 1990, vol 7, pp 1–12

Wave Propagation, Activity and Frequency Selectivity in the Cochlea

E. de Boer[1]

Academic Medical Centre, Amsterdam, The Netherlands

In several respects the field of modelling the mechanics of the cochlea
is far developed but it seems that new experimental results perpetually
make existing models obsolete. In the first era of modelling [1], only the
stiffness and resistance of the organ of Corti (OC) were invoked, and not
the mass, because available mechanical data showed a very low degree of
mechanical tuning. Later on, the mass had to be included because, as time
went on, the observed mechanical tuning appeared to be sharper and
sharper. More recent mechanical data show responses which – at low levels
of stimulation – are almost as selective as neural responses [2]. In addition,
the types of nonlinearity are similar.

It has turned out to be well-nigh impossible to simulate these sharply
tuned responses by a simple type of cochlear model. Several authors
devised a way out of this difficulty by postulating that the cochlea is
mechanically active [3–5]. This means it is assumed that certain structures
of the cochlea are capable of amplifying the ongoing cochlear wave by
actively setting the basilar membrane in oscillatory motion. For a given
location along the length of the cochlea, activity would mostly be needed
over a limited range of frequencies, namely, over approximately the half
octave below the best frequency for that location. Activity is not primarily
needed to sharpen the response peak. By resorting to a secondary resonator
Neely and Kim [5] could meet these requirements and achieve a quite
realistic response in their active model.

[1] The author thanks Drs. Jonathan Ashmore, David Kemp, Max Viergever and Diek
Duifhuis for their contributions in stimulating discussions and their constructive criti-
cism.

One major problem inherent in active models is stability. Activity is in many respects equivalent to positive feedback. Hence a central question to ask is: How does the system, in view of anatomical and physiological variations, remain stable? A satisfactory solution to this problem has not yet been found. Nonlinearity will probably play an important part. Only recently [6] have the required computational tools been developed far enough to make really efficient computations on active and nonlinear models possible.

The alternative, an active and nonlinear model that is unstable at every location (a 'pan-active' or 'pan-ergic' model) has only been studied superficially [7]. Another alternative, a complex passive model that is based on the anatomical differences between pars tecta and pars pectinata of the basilar membrane (BM; the OHCAP model) [see ref. 8] is at present also in an early stage of development.

There is one feature that is often overlooked in the formulation of active cochlear models. It has never been shown that a structure that is imbedded in the OC would be able to impart acoustical energy to the BM. This question has become extremely important since it has been discovered that outer hair cells (OHCs) can show fast motile responses [9, 10]. It would be quite attractive to assume that OHCs exert forces on the BM, either by lengthening or shortening [9] or by bending [10]. That OHCs play a crucial part in controlling cochlear frequency selectivity has been known for a long time [11] which makes this assumption even more attractive.

In the present paper the question of whether or not structures like OHCs can amplify cochlear waves is studied by way of a simple model in which OHCs are able to induce relative movements in the OC. It will be shown that the influence of the OHCs on wave propagation in the cochlea is, in general, so small that it is unlikely that the motile responses observed can be the source of cochlear activity. Nor can these responses be expected to contribute to frequency selectivity.

In the final section a few speculations are described on the 'real' source of cochlear activity and the possible function of outer hair cell motility.

The Sandwich Model

In an active model amplification of the cochlear wave would mostly be needed in the part of the cochlea where the cochlear wave is travelling towards its peak, i.e., in a region where the wavelength is larger than the cross-sectional diameter of the cochlear channel. Hence, at the present stage of development, it is sufficient to consider a long-wave model. Needless to say that for analytical treatment, a linear model is mandatory; the model should simulate the behaviour of the real cochlea at the lowest levels of stimulation.

The most essential point to envisage now is that the OC should be deformable in some sense. It has been decided to choose a model where the OC deformations occur in the form of (local) area changes. That other types of OC deformation have a very small interaction with the cochlear wave is shown elsewhere. The OC is modelled in the form of two parallel flexible membranes stretched from one wall of the cochlea to the other. One membrane is called the reticular lamina (RL), the other is the BM. In between these two lies the OC, to be modelled as a fluid channel. Relative movements of the membranes are possible only in the sense that they approach each other or move farther apart, they remain parallel to each other. Such movements can occur, e.g., by length changes of OHCs.

A cross-section of the model (perpendicular to the x-axis) is shown by figure 1. This figure also shows the symbols to be used further on, in particular, the various pressures involved and their relations to the two membrane velocities. The impedances Z_{BM} and Z_{RL} yield the pressures with which the membranes BM and RL oppose movements. In addition, the impedance Z_{OC} represents the way the two membranes are attached to each other, this impedance produces the pressure against relative movements. Finally, an impedance of a different kind, the transfer impedance Z_{HC}, is invoked to represent the pressure p_{HC} exerted by the OHCs. In all these physical quantities (pressures, velocities and impedances), the independent variable x on which they depend is left out; this will be done throughout this paper at those places where it does not give rise to confusion.

One essential point to take into account is that a force exerted on one membrane by a structure inside the OC has as its counterpart the *same* force in *opposite* direction at the other membrane. Such balanced forces will, by themselves, not be able to move the OC as a whole. This reasoning applies to the pressure p_{OC} associated with Z_{OC} as well as to the pressure p_{HC} generated by the OHCs (see fig. 1). Only when the system is brought into an unbalanced state can we expect the pressure p_{HC} to become effective.

In classical models, the OC is assumed to be incompressible at every location. In the present model it can be compressed or dilated locally but a constriction at one location will be offset by dilatations at other locations,

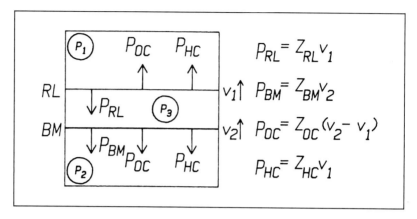

Fig. 1. Cross-section of sandwich model.

the total volume stays constant. In fact, the OC is a fluid-filled channel with two flexible walls (channel 3), it is located between two other channels (1 and 2) and closed at both ends. The name 'sandwich model' should be self-evident by now.

In order to analyze the dynamics of channel 3 let us first describe that of channel 1, a channel with one flexible wall. Consider a short section of channel 1 with length dx and cross-sectional area A_1. The net outward flux of fluid is $-A_1 dx (du_1/dx)$ where u_1 is the longitudinal velocity of the fluid. The net fluid flux inward is due to movements of the RL, expressed by the velocity v_1. Equality of influx and outflux leads to the equation

$$b_1 v_1 = -A_1 \frac{du_1}{dx},$$

(1)

where b_1 is the (effective) width of the RL. When the pressure in channel 1 is p_1, the gradient of p_1 produces acceleration of the fluid (in the x-direction). By combining this property with equation (1), the fluid equation for channel 1 is obtained as:

$$\frac{d^2p_1}{dx^2} = -i \omega D_0 \frac{v_1}{h_1},$$

(2)

where D_0 is the fluid density, and $h_1 = A_1/b_1$, the effective height of the channel. The quantity ω is the radian frequency $2\pi f$, we are tacitly considering sinusoidal movements at one frequency, f. This reasoning closely parallels the first phase in the derivation of the classical long-wave model equation. In the same way the equation for channel 2 can be derived (note the different sign):

$$\frac{d^2p_2}{dx^2} = +i\,\omega\,D_0\,\frac{v_2}{h_2}, \tag{3}$$

where v_2 is the velocity of the BM, p_2 is the pressure in channel 2 and h_2 the effective height of that channel. We will henceforth make $h_2 = h_1$ and use the symbol h for both. It is now straightforward to write down the fluid equation for channel 3:

$$\frac{d^2p_3}{dx^2} = -i\,\omega\,D_3\,\frac{v_2 - v_1}{h_3}, \tag{4}$$

where p_3 is the pressure and h_3 the effective height. The quantity D_3, the fluid density, is not a real number because the fluid in channel 3 is moving with considerable friction. For simplicity, we will make D_3 equal to D_0 and absorb the imaginary component of D_3 into h_3 by making h_3 complex. To represent frictional losses, the imaginary part of h_3 should be negative. Note that the magnitude of h_3 is smaller than h. This might amount to a factor of 4–10 (or even larger) as judged from cross-sectional pictures.

Substitution of equations (2) and (3) into equation (4) gives:

$$\frac{d^2p_3}{dx^2} = -\frac{h}{h_3}\frac{d^2}{dx^2}(p_1 + p_2), \tag{5}$$

and this equation can directly be integrated twice to yield

$$p_3 = -\frac{h}{h_3}(p_1 + p_2). \tag{6}$$

When both p_1 and p_2 are zero for all x, p_3 must be identically zero too. Therefore, the two integration constants are zero.

At this point it is possible to discern a far-reaching consequence of the structure studied. Assume, for a moment, that a mode of propagation is possible in which p_2 equals minus p_1 for every x. Equation (6) then tells us that p_3 will be zero, there will be no compressional pressure acting on the OC. Conversely, the two membranes will move with the same velocities and there will be no influence of the OHCs whatsoever. A coupling from OHCs to fluid waves (or vice versa) can exist only when $p_1 + p_2$ is not equal to zero, or: p_2/p_1 is not equal to -1. We will come back to this point later.

The final step in the derivation of the model equations is to relate the pressures to the velocities v_1 and v_2 via the various impedances. On the RL act the fluid pressures p_1 and p_3, and these are balanced by the restoring pressures p_{RL}, p_{OC} and p_{HC}. As can be seen from the formulae in figure 1, it is assumed that p_{HC} is controlled only by movements of the RL, via the transfer impedance Z_{HC}. The equilibrium of pressures at the level of the RL is described by the following relation:

$$p_1 - Z_{HC}\, v_1 + Z_{RL}\, v_1 - Z_{OC}\, (v_2 - v_1) - p_3 = 0. \tag{7}$$

In an entirely analogous way, the following relation holds true for the BM:

$$p_2 - Z_{HC}\, v_1 - Z_{BM}\, v_2 - Z_{OC}\, (v_2 - v_1) - p_3 = 0. \tag{8}$$

It is simple to solve these equations for v_1 and v_2 and to eliminate the velocities with equations (2) and (3). This procedure gives rise to the following model equations:

$$\frac{(b\, c - a\, d)}{E}\,\frac{d^2 p_1}{dx^2} = - d\, p_1 + b\, p_2 + (d - b)\, p_3 \tag{9}$$

$$\frac{(b\, c - a\, d)}{E}\,\frac{d^2 p_2}{dx^2} = - c\, p_1 + a\, p_2 + (c - a)\, p_3, \tag{10}$$

where $E = i\, \omega\, D_0/h$, $a = Z_{OC} + Z_{RL} - Z_{HC}$, $b = - Z_{OC}$, $c = Z_{OC} - Z_{HC}$ and $d = - (Z_{OC} + Z_{BM})$. When equation (6) is substituted for p_3, the result is a set of two coupled second-order differential equations in p_1 and p_2. Together, the equations describe a fourth-order system.

The boundary conditions are determined as follows. At $x = 0$ the velocity of the stapes determines the longitudinal fluid velocity $u_1(0)$. Since channel 3 is closed at this point, and the total fluid volume must be constant, the fluid velocity $u_2(0)$ in channel 2 must be equal to $-u_1(0)$. At the helicotrema end of the model we let both fluid velocities go to zero.

The Physics of the Sandwich Model

From the foregoing it is evident that in the sandwich model the topics of wave reflection (important for cochlear echoes and spontaneous emissions) and resonance are more complicated than in the simple classical two-channel model. Therefore, we leave reflections and resonance out of the present discussion and continue on wave propagation and its relation to activity. Assume now, for a moment, that all model parameters are constant. The model equations then describe a linear fourth-order system with constant coefficients. Hence, there are four eigen values and four eigen functions of the form exp $(-i \, k_j \, x)$ with $j = 1,..,4$. If we assume that there are no reflections, only two of these will be relevant. The actual solution will be a linear superposition of the two eigen functions.

For each eigen value k_j the pressures p_2 and p_1 will have a constant ratio α_j, defined as $\alpha_j = p_2/p_1$, with $j = 1,2$. Take now the case in which we expect that the influence of OHC mechanics is largest, namely, the one where both Z_{RL} and Z_{OC} are zero. Furthermore, use the symbol η for Z_{HC}/Z_{BM}. It is a matter of algebra to show that the two values of α_j are the roots of the following quadratic equation in α:

$$\frac{h}{h_3} \alpha_2 + (\frac{h}{h_3} + 1 + \eta) \, \alpha - \eta = 0 \tag{11}$$

See the Appendix for the derivation. For small $|\eta|$ one of the roots is:

$$\alpha_1 = \frac{\eta}{(h/h_3) + 1 + \eta}. \tag{12}$$

This value represents a solution of the model equations in which the RL is vibrating with a much larger amplitude than the BM. It is as if a wave is travelling along the length of the OC, riding as it were 'on the back' of the BM. The propagation speed is mainly determined by the mechanical properties of the OC, namely, by Z_{HC}. This slowly propagating wave (note: $|Z_{HC}|$ is assumed to be smaller than $|Z_{BM}|$) may be rapidly damped out, especially since h_3 contains a substantial imaginary component representing fluid friction inside the OC.

The other root α_2 is, again for small $|\eta|$, given by:

$$\alpha_2 = -1 - \frac{1 + \eta}{(h/h_3)} - \frac{\eta}{(h/h_3) + 1 + \eta}, \tag{13}$$

a value that is not far from -1. As outlined earlier: for such a value there will be an inherently weak coupling between OHC mechanics and the cochlear wave. Substitution of this value in equation (10) produces a differential equation not unlike that of the long-wave model. The BM impedance occurs in this equation in the form $\frac{1}{2} (Z_{BM} + Z_{HC}/m)$, where m is a reduction factor. The factor of $\frac{1}{2}$ occurs also in the classical long-wave model as a result of α being exactly equal to -1. The value of m can be estimated from an approximate calculation in which it is assumed that $|h/h_3|$ is large ($>> 1$) and $|\eta|$ is small ($<< 1$). The result is:

$$m = \frac{2 (1 + d_3)^2}{1 + 2 \, d_3} \tag{14}$$

with $d_3 = h/h_3$. For large $|d_3|$ this is approximately equal to d_3. And as explained earlier, the magnitude of this parameter would be fairly large, 4–10. It should be clear that for larger values of $|\eta|$ the same property obtains: the influence of the HC impedance is reduced by a factor of the order of (h/h_3). This elaboration serves to illustrate how small the influence of OHC mechanics on the cochlear waves really is. If OHC motility were the source of cochlear activity, the OHCs should be able to produce pressures much larger than those produced by the BM. It should be noted, furthermore, that the 'natural' excitation of the cochlea, where the fluid velocities in the two main channels are of opposite sign, tends to produce a situation where the eigen function with this value of α is present with the largest amplitude.

Let us finally consider the case where the two values α_1 and α_2 are nearly equal (and where the two eigen values are nearly equal too). This occurs when the discriminant of eq. (11) is (nearly) zero. It follows that $|\eta|$ has to be of the order of $|h/h_3|$ which again means that the OHCs have to provide pressures that are much larger than those produced by Z_{BM}. This is an equally unlikely situation as the one described before.

Generalization, Conclusions

One further step is needed to make the argument more general and conclusive. Starting from the constant-parameter case, take the model parameters as slowly varying functions of x. Each eigen function then turns into a characteristic 'mode' in which wave amplitude, propagation speed and α vary, also slowly, with x. As a matter of fact, the principal mode in the classical long-wave model is nothing else than the LG (or WKB)[2] solution: see, e.g. reference 12. For the present model the LG method can be applied to both modes, or, more precisely: to the sum of the modes. It has been shown (to be published elsewhere) that each of the modes develops in its own way, independent of the other one. A transfer of acoustical energy from one mode to the other does not occur. The derivation is valid for any fourth-order system with any combination of parameters. The only condition for the entire derivation is that the model parameters vary so slowly with x that no internal reflection occurs.

---

[2] LG = Liouville-Green; WKB = Wentzel-Kramers-Brillouin; abbreviations for an asymptotic method for solving second-order differential equations.

It follows that the conclusion for the constant-parameter case can be extended to the case of a general model with varying coefficients. Again, the influence of forces exerted by the OHCs is reduced by a substantial factor. And the mode in which such a small influence occurs is dominant because it has a relatively large amplitude.

Considerable experience with the sandwich model failed to reveal any property that could not be explained in terms of what has been said above. The system appears to have an innate tendency to let the first mode decay quickly. The second mode remains and in this mode the influence of the OHCs is reduced by a substantial factor. (An exception occurs when the parameters are such that local reflections occur; such cases are to be excluded, however, because the associated phase response is contrary to what is found experimentally.) Apparently, the coupling between hair cells and the macromechanics of the cochlea is so weak that a cochlear amplification system based upon it is quite inefficient. It is, therefore, unlikely that OHCs can serve as the source of acoustical energy in the structure of an active model.

This conclusion has been derived, of course, only for the specific type of OHC mechanics envisaged here, i.e. a system in which OHCs exert forces that can move RL and BM farther apart or closer together. As mentioned earlier, the coupling between OHCs and cochlear waves is also very small when the cross-sectional area remains constant during the deformations of the OC (the solution for this situation requires the full machinery of three-dimensional modelling). In addition, the conclusion remains the same when it is taken into account that the fluid pressure p_1 acts only on a part of the RL. The conclusion also remains true in the case where, due to the phalangeal processes of Deiters' cells, each OHC produces distributed forces at two or more locations. The possibility remains open, of course, that the motile responses express themselves in cochlear mechanics in a way that differs fundamentally from the one treated here and the variations mentioned above. On physical grounds, it is to be doubted whether this could be true. Therefore, the conclusion of the preceding paragraphs is fairly general and should be taken quite literally. It is only subject to the restriction of long-wave behaviour.

In some respects the formulation of active cochlear models now seems to have been a more or less abstract exercise. This would apply to, e.g., the model of Neely and Kim [5] in which it is assumed that the active force acts directly and only on the BM. Furthermore, the principle of undamping by OHC motility as demonstrated by Geissler [13] would also be incorrect.

In that paper the dual action of the forces exerted by the OHCs is duly taken into account but the effect of the two channel pressures p_1 and p_2 is left out.

Speculations

If we accept the view that some sort of activity is present in the cochlea, we still have to guess at its mode of generation. If the OHC motile response (in the form of oscillatory length changes or tilts) is not the source of activity, what else is it? It could be that the cilia of the OHCs move actively. For instance, Gitter and Zenner [10] state that the cilia partake in the motile responses of the OHCs. On second thought, this possibility is to be ruled out. The stiffness of the cilia is probably much smaller than that of the BM (compare, for instance, the parameters k_3 and k_1 in Neely and Kim [5], these differ by two orders of magnitude). It is highly unlikely that by active means the cilia would produce forces that are so much larger than reactions to displacements of the basilar membrane. The same argument can be used for any other mechanism that would transmit the generated force via the cilia.

Let us consider the issue from a completely different side. What would be the primary store of energy from which the active force is derived? Various possibilities present themselves: electrical energy, chemical energy, etc. That it could be mechanical energy has not yet been discussed in the literature. Assume that a state of steady tension (or stress) is maintained in the cochlear partition, and that in each period of a vibration, over a limited phase region, a tiny part of that tension is released and given to the OC so as to enhance its motion. That mechanism would be analogous to the action of the bow on a violin string. A question to be asked to physiologists: would such a process be possible? The source of the tension could reside in the d.c. fluid pressure in the scala media, or the tension could directly be produced by the tectorial membrane. The modulation of the tension is done by a mechanism (a version working on the molecular level is possible [14]) that could be located between the tectorial membrane and the reticular lamina, a real 'white area' in the map of the cochlea. The OHCs might well play a part in it. All this is pure speculation, of course.

A remaining question is: what is the function of the lengthening and shortening of OHCs as observed in vitro? Within the same realm of speculation it can be surmised that length changes of OHCs may control sensitivity and selectivity of the cochlea. When the cells increase their length, the (outer part of the) RL is lifted, and the underside of the tectorial membrane is moved away from the OC. This results in a larger distance between the tectorial membrane and the inner hair cells and weaker stimulation of the latter cells' cilia. By the same token, the OHCs have shifted their working point so that the amount of activity decreases. Both effects would serve to explain some of the effects of efferent stimulation as they have been reported, e.g. by Wiederhold and Kiang [15].

Appendix

Write the model equations as:

$$\frac{d^2 p_1}{dx^2} = A\, p_1 + B\, p_2 \tag{A-1}$$

$$\frac{d^2 p_2}{dx^2} = C\, p_1 + D\, p_2 \tag{A-2}$$

and try to solve these equations under the assumption $p_2 = \alpha\, p_1$. For the proper value of α the two equations must be equivalent and their coefficients proportional to one another. Substitution of $p_2 = \alpha\, p_1$ gives the following form to this condition:

$$\frac{C + \alpha\, D}{A + \alpha\, B} = \alpha \tag{A-3}$$

from which the quadratic equation for α turns out to be:

$$B\, \alpha^2 + (A - D)\, \alpha - C = 0. \tag{A-4}$$

References

1 Zwislocki J: Theorie der Schneckenmechanik: Qualitative und quantitative Analyse. Acta Oto-Laryngol 1948;suppl 72.
2 Sellick PM, Patuzzi R, Johnstone BM: Measurement of basilar membrane motion in the guinea pig using the Mössbauer technique. J Acoust Soc Am 1982;72:131–141.
3 Kim DO, Neely ST, Molnar CE, et al: An active cochlear model with negative damping in the partition: Comparison with Rhode's ante- and post-mortem observations; in Brink G vd, Bilsen FA (eds): Psychophysical, Physiological and Behavioural Studies in Hearing. Delft, Delft University Press, 1980, pp 7–14.
4 Boer E de: No sharpening? A challenge for cochlear mechanics. J Acoust Soc Am 1983;73:567–573.
5 Neely ST, Kim DO: A model for active elements in cochlear biomechanics. J Acoust Soc Am 1986;79:1472–1480.

6　Diependaal RJ, Viergever MA: Nonlinear and active two-dimensional cochlear models: Time-domain solution. J Acoust Soc Am. In press, 1989.

7　Duifhuis H, Hoogstraten HW, Netten SM van, et al: Modelling the cochlear partition with coupled Van der Pol oscillators; in Allen JB, Hall JL, Hubbard AE, et al (eds): Peripheral Auditory Mechanisms. New York, Springer, 1985, pp 290–297.

8　Kolston PJ, Viergever MA, Boer E de, et al: Realistic mechanical tuning in a micromechanical model. J Acoust Soc Am 1989;86:133–140.

9　Ashmore JF: A fast motile response in guinea-pig outer hair cells: the cellular basis of the cochlear amplifier. J Physiol (Lond) 1987;388:323–347.

10　Gitter AH, Zenner H-P: Auditory transduction steps in single inner and outer hair cells: in Duifhuis H, Horst JW, Wit HP (eds): Basic Issues in Hearing. London, Academic Press, 1988, pp 33–39.

11　Dallos P, Harris D: Properties of auditory-nerve responses in absence of outer hair cells. J Neurophys 1978;41:365–383.

12　Boer E de, Viergever MA: Wave propagation and dispersion in the cochlea. Hearing Res 1984;13:101–112.

13　Geissler CD: A model of the effect of outer hair cell motility on cochlear vibrations. Hearing Res 1986;24:125–131.

14　Bialek W, Wit HP: Quantum limits to oscillator stability: Theory and experiments on acoustic emissions from the human ear. Phys Lett D 1984;104A:173–177.

15　Wiederhold ML, Kiang NY-S: Effects of electrical stimulation of the crossed olivo-cochlear bundle on single auditory-nerve fibers in the cat. J Acoust Soc Am 1970;48: 950–965.

Prof. Dr. E. de Boer, D2-210, Academic Medical Centre, Meibergdreef 9, NL–1105 AZ Amsterdam (The Netherlands)

Grandori F, Cianfrone G, Kemp DT (eds): Cochlear Mechanisms and
Otoacoustic Emissions. Adv Audiol. Basel, Karger, 1990, vol 7, pp 13–26

Cellular Mechanical Responses in the Cochlea[1]

Shyam M. Khanna[a], *Mats Ulfendahl*[b], *Åke Flock*[b]

[a] College of Physicians & Surgeons, Columbia University, New York, N.Y., USA;
[b] Department of Physiology II, Karolinska Institute, Stockholm, Sweden

Until recently it was generally accepted that the basilar membrane
mechanics dominated the tuning characteristics of the inner ear. Khanna
and colleagues concluded from their vibration measurements and histological evaluation of cat cochleas that the vulnerable sharply tuned component of the mechanical response seen at the basilar membrane originated at
the outer hair cells [1–9]. In order to demonstrate this concept directly it
was necessary to measure the mechanical cellular response without appreciably disturbing the cochlea. A special instrument consisting of a combination of an optical sectioning microscope [10] and heterodyne interferometer [11] was built to carry out such measurements. Details of this
instrument are described elsewhere [12]. Using this instrument, cellular
vibrations were measured in the guinea pig organ of Corti in response to
acoustic signals applied to the ear. The preparation of the temporal bone,
opening of the cochlea and the basic measurement technique have been
described in a companion paper [13].

Responses of the Outer Hair Cells and the Basilar Membrane

A comparison of the vibration amplitude measured at the outer hair
cells and the basilar membrane in the fourth turn was made earlier [14]. In
one experiment, the frequency maximum of the tuned response of an outer

[1] This research was supported by Program Project Grant NS22334 from NIDCD.

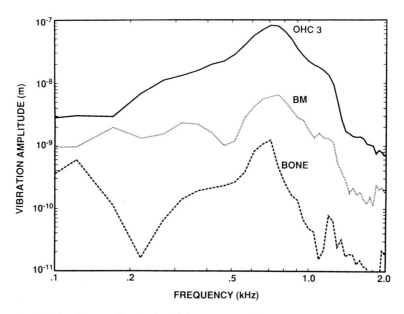

Fig. 1. Vibration amplitude of a third row outer hair cell measured in the third turn of the cochlea (solid line). The frequency of maximum vibration is 710 Hz and the vibration amplitude is 8.0×10^{-8} m. The basilar membrane response is shown by the dotted line. A resonance is seen in the 700-Hz frequency region. The vibration amplitude of the membrane at 710 Hz is 6.0×10^{-9} m. It is lower by a factor of 13 than the outer hair cell response. The vibration amplitude of the bony shell is shown by the dashed line. This response too shows a resonance in the 700-Hz frequency region. The vibration amplitude of the bony shell at 710 Hz is 1.2×10^{-9} m.

hair cell was 220 Hz. The response of the basilar membrane measured close to the outer hair cell showed no tuning and its response amplitude at 220 Hz was less than 1/200 of that measured at the outer hair cell. These observations show that in our preparation the mechanical tuning of the outer hair cells is independent of the basilar membrane mechanical response. The vibrations of an outer hair cell, the basilar membrane and the bony shell measured in the third turn of the cochlea for an input sound pressure level of approximately 94 dB are shown in figure 1. The outer hair cell response shows a maximum at 710 Hz with an amplitude of 0.8×10^{-7} m. The basilar membrane response shows tuning in the third turn but its amplitude is lower (6×10^{-9} m). A resonance is also seen in the bony shell. Its vibration amplitude is still lower (1.2×10^{-9} m). These differences in response amplitude suggest that the vibration of outer hair cells induces vibrations in the basilar membrane as well as in the bony shell adjacent to that region.

Hair Cell Tuning along the Length of the Cochlea

The tuning characteristics of outer hair cells change when measurements are made at different longitudinal positions in the cochlea. Comparison of the responses measured in the fourth turn with responses measured in the third turn (fig. 2) shows the following: (1) frequency of maximal response is higher for the third turn; (2) the tuning curve is sharper for the response measured in the third turn; (3) the high-frequency slope of the response is steeper than the low frequency slope in the third turn, and (4) phase delay is smaller in the third turn. Each of these characteristics has been observed in the tuning characteristics recorded from individual fibers of the eighth nerve [15].

Cellular Vibrations along a Radial Track

How are the vibrations of different cells related to one another? To answer this question, tuning characteristics were measured in the region of Hensen's cells, in each row of OHC, in the pillar cells and in the inner hair cells at the level of the reticular lamina along a radial track. Only vibrations perpendicular to the plane of reticular lamina were measured. The vibration amplitude was highest at the outer edge of Hensen's cells and decreased when moving inward toward the inner hair cells. The decrease was approximately linear with distance from the outer Hensen cell edge. This suggested that the reticular lamina vibrated like a hinged stiff plate.

The results were dependent on the condition of the cochlea. In cochleae with damage, the tuning characteristics of all cell types are quite similar in shape, and are lower in amplitude (fig. 3a). When the cochlea is in good condition, tuning characteristics measured at each cellular location are similar, but there are differences near and above the frequency of maximal response (fig. 3b). Vibration amplitude decreases nonlinearly with distance, and the slope of the curve changes at each cellular region

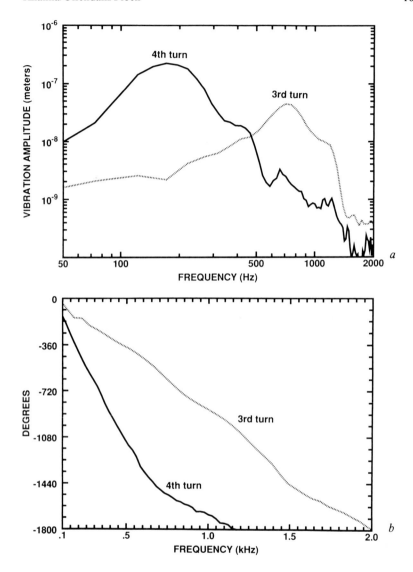

Fig. 2. a Solid line shows the vibration amplitude of an outer hair cell measured as a function of frequency in the fourth cochlear turn approximately 18.5 mm from the base. Dotted line shows a similar response measured in the third cochlear turn approximately 13.5 mm from the base. The tuning of the two responses shows important differences: (1) the frequency of maximum is higher for the basal measurement (708 Hz at 13.5 mm,

(fig. 3c). The slope increases in steepness at each outer hair cell location, reaching a maximum near the third row of outer hair cells. The slope is shallow in the region of Hensen's cells. In a damaged cochlea the increase in amplitude from the first row of outer hair cells to the second, and from the second to the third is a factor of 1.2. In a relatively undamaged co-chlea, the increase is a factor of 2.0. This large increase in the outer hair cell region and the lack of a similar effect in the region of the inner hair cells or Hensen's cells suggests that the outer hair cells drive the reticular lamina. Each hair cell acts to increase the vibration amplitude of the reti-cular lamina. The amount of increase is dependent on the condition of the outer hair cells.

Mode of Vibration of the Reticular Lamina

The vibration amplitude of a Hensen cell was measured in different directions by changing the angle between the measuring beam and the lamina (fig. 4). Knowledge of vibration amplitudes, vibration phases and angles of measurement (fig. 5) allows precise determination of the angle at which maximum vibration occurs, and the axis of the maximum vibration is very close to the axis of the outer hair cells. The vibrations of the underlying basilar membrane are small in relation to the vibrations of the outer hair cells and the reticular lamina. It appears that the outer hair cells themselves must be changing their length and carrying the reticular lamina with them. The motility of isolated outer hair cells under an external elec-tric field has been demonstrated [16]. The isolated outer hair cells also display fast motility when subjected to an external alternating field [17]. The shortening of the isolated outer hair cells has also been demonstrated under the influence of substances that induce contraction of muscle cells [18–20], and with acoustic stimulation [21].

---

171 Hz at 18.5 mm); (2) the basal response is more sharply tuned; and (3) the basal response is more asymmetrical. The high frequency slope is much steeper than the low frequency slope. *b* The vibration phase is measured with respect to the phase of the electrical input applied to the acoustic transducer. The phase delay is much higher in the apical 18.5 mm position than in the 13.5 mm position.

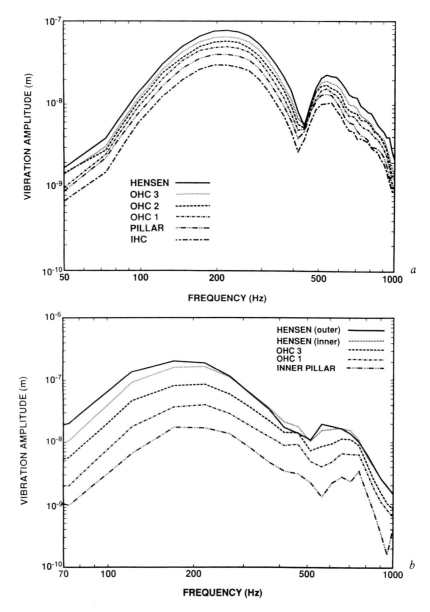

Fig. 3. a Vibration amplitude measured as a function of frequency at six positions along a radial track. (1) Hensen's cell (solid line), (2) outer hair cell third row (dotted line), (3) outer hair cell second row (dashed line), (4) outer hair cell first row (dashed dotted line), (5) middle of outer pillar head (dashed triple dotted line), and (6) inner hair cell (line with long and short dashes). The tuning curves look similar in shape for all the

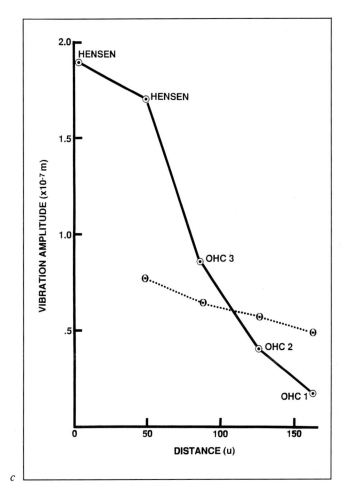

c

points measured, but their magnitude decreases from the Hensen's cells towards the inner
hair cells. *b* Vibration amplitude measured as a function of frequency in a second cochlea
at five positions along a linear track: (1) the outer edge of Hensen's cells (solid line),
(2) the inner edge of Hensen's cells (dotted line); (3) the second row of outer hair cells
(dashed line); (4) the first row of outer hair cells (dashed dotted line), and (5) the inner
hair cells (dashed triple dotted line). The shape of the tuning curves is generally similar
but the magnitude of the response decreases as the point of measurement moves inwards
from Hensen's cells toward the inner hair cells. Differences in the response are more
pronounced in the frequency region following the peak, in the region of the first minimum, and in the region of the second peak. *c* Cellular vibration amplitude as a function
of distance from the outer edge of Hensen's cells. The amplitude at 219 Hz from *(a)* is
shown with a dotted line and that from *(b)* is shown with a solid line. The slope of the
dotted line is nearly linear but that of the solid line changes with cellular position.

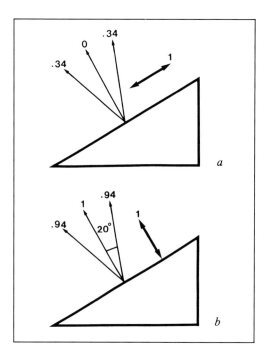

Fig. 4. Two possible modes of vibration of the reticular lamina. *a* The vibration is purely along the plane of the lamina. In this case the component measured in the normal viewing direction will be zero. A 10° offset of the measuring axis will increase it to 0.17 and a 20°offset to 0.34. *b* The vibration is only along the normal viewing direction and its magnitude is unity. A change in viewing direction by 10° will decrease it to 0.98. A change of 20° will decrease it to 0.94.

Waveform and Spectrum of Response

The velocity waveforms (fig. 6) of the vibrations of outer hair cells are distorted, mainly at high input levels of sound (94 dB SPL). The distortion is more pronounced when the test frequency is near the frequency of maximal response or just above it, and the degree of distortion is highly dependent on the condition of the cochlea. In damaged cochleae the response is nearly sinusoidal, and the waveform does not look distorted. The spectrum of the response, obtained by Fourier transformation of the velocity waveform, consists of components at the fundamental frequency and its harmonics (fig. 6). The magnitude of the spectral components and the number

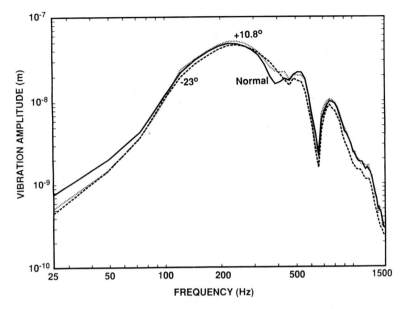

Fig. 5. Vibration amplitude of a Hensen's cell measured as a function of frequency at three different viewing angles. Normal viewing angle (solid line) perpendicular to the reticular lamina, at +10.8° (dotted line) and at −23° (dashed line). Changes with viewing angle are small but maximum amplitude is observed at +10.8° indicating that the vibration axis is tilted with respect to the normal to the reticular lamina by +10.8°.

of components seen in the spectrum vary with the signal level and with the test frequency. Both the magnitude of the components and their number increase with signal level. At any given signal level, the spectral magnitude and number is highest at the frequency of maximal response, or just above this frequency. At frequencies far below and far above the frequency of maximal response, the spectral magnitude of the harmonics is significantly lower, and very few spectral components are seen.

Harmonic Tuning Curves

The amplitude of response can be plotted for any selected harmonic as a function of stimulus frequency in the same way that responses for the fundamental are plotted. In some experiments, the second harmonic com-

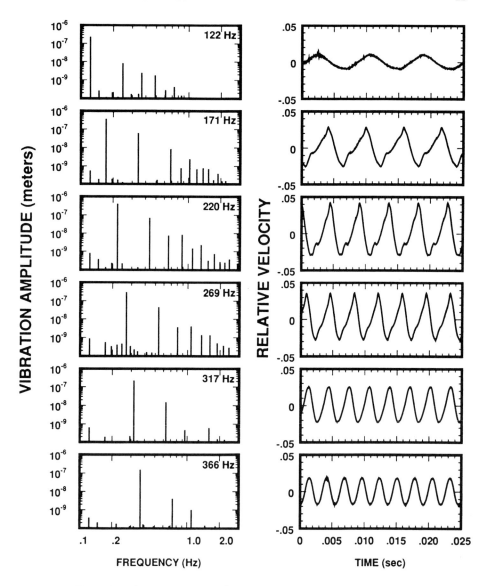

Fig. 6. Waveforms and spectra of vibratory responses of a first row outer hair cell in the apical turn of a guinea pig cochlea in response to a 5.0-volt peak sinusoidal signal applied to the acoustic transducer. The relative velocity amplitude is shown on the right as a function of time for six test frequencies. The waveform shows distortion during the negative half of the cycle at 122 Hz. The distortion affects both negative and positive portions of the cycle at 171 Hz and the wave shape looks like a sawtooth. The distortion increases at 220 Hz and a clipping of the negative peak is seen. The distortion decreases at

ponent of the response shows much sharper tuning than the fundamental (fig. 7). The input frequency that produces a maximal response may be the same for the two curves in the beginning, but the shape and frequency of the maximal response of the second harmonic tuning curve change rapidly in any given experiment. The response amplitude is reduced with time as the tuning curve broadens, often asymmetrically. In comparison, changes seen in the tuning curves for the fundamental component of the response are small. Furthermore, changes in the shapes of the harmonic tuning curves with time are not related to changes seen in the tuning curves for the fundamental component of the response. These observations suggest that the harmonic and fundamental components have different origins.

Discussion

The measurements of cellular vibration give us a new concept of how the auditory stimulus is detected by the sensory receptors. It is quite different from the classical concept in which the stereocilia bend passively in response to the mechanical stimuli shaped by the basilar membrane mechanical characteristics.

The outer hair cells display a sharply tuned response, in which their cell bodies change their length in an oscillating manner. The mechanism by which the hydrodynamic stimulus is sensed by the outer hair cells is not clear at present. The tuning is an active process, as suggested in our companion paper [13]. The motor response of the outer hair cells at the fundamental frequency appears to be linear up to the levels tested in our experiments, and drives the reticular lamina. The tall rows of the bundle are embedded in the tectorial membrane. The tectorial membrane, stereocilia bundle and cuticular plate thus provide a dynamic load to the outer hair

---

269 Hz where the sawtooth shape reappears. At higher frequencies (317 and 366 Hz) the distortion reduces further and the waveform becomes nearly sinusoidal. The spectrum corresponding to the velocity waveforms is shown on the left, but instead of showing velocity amplitude, the spectrum is calculated to show the absolute vibration amplitude at the signal frequency and its harmonics. The number of harmonics in the spectrum increases from five at 122 Hz to ten at 220 Hz. The number is eight at 269 Hz, three at 317 Hz and two at 366 Hz.

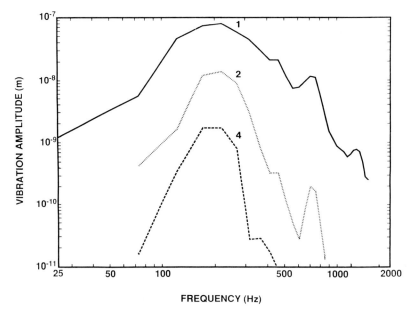

Fig. 7. Vibration amplitude of an outer hair cell (1st row) as a function of frequency. The solid line shows the fundamental component, the dotted line the second harmonic component and the dashed line the fourth harmonic component. All measurements were made with maximum driving voltage (5-volt peak) applied to the acoustic transducer. Amplitudes shown have been linearly scaled for 1 V. The shape of the second and fourth harmonic tuning curves is quite different from that of the fundamental. The peaks of all the curves are located at the same frequency (220 Hz), but the second and the fourth harmonic curves show much sharper tuning.

cell motion. The directional asymmetry in the stereocilia stiffness [22] may introduce nonlinearity in the mechanical response and this is seen in the waveform and spectrum of the measured response. The vibration of the reticular lamina is mechanically coupled to the apical end of the inner hair cell, therefore providing an inertial stimulation of the cuticular plate and the attached stereocilia bundle. There may be other pathways of mechanical coupling to the inner hair cell through the tectorial membrane.

Measurement of cellular mechanics gives us a picture of the inner ear as a complex dynamic system whose mechanical characteristics are determined at a cellular level and are dependent on the metabolic conditions and control of the central nervous system. Understanding the details of the function will be an exciting challenge for auditory science.

References

1 Khanna SM, Leonard DGB: Basilar membrane tuning in the cat cochlea. Science 1982;215:305–306.
2 Khanna SM, Leonard DGB: An interpretation of the sharp tuning of the basilar membrane mechanical response; in de Boer E, Viergever MA (eds): Mechanics of Hearing. Delft, Delft University Press, 1983, pp 177–181.
3 Khanna SM, Leonard DGB: Measurement of basilar membrane vibrations and evaluation of cochlear condition. Hear Res 1986;23:37–53.
4 Khanna SM, Leonard DGB: Relationship between basilar membrane tuning and hair cell condition. Hear Res 1986;23:55–70.
5 Leonard DGB, Khanna SM: Histological evaluation of damage in cat cochlea used for measurement of basilar membrane mechanics. J Acoust Soc Am 1984;75:515–527.
6 Khanna SM: Interpretation of the sharply tuned basilar membrane response observed in the cat cochlea; in Fay RR, Gourevitch G (eds): Hearing and Other Senses. Groton, Amphora Press, 1983; pp 65–86.
7 Khanna SM: Inner ear function based on the mechanical tuning of the hair cell; in Berlin C (ed): Hearing Research: Recent Advances. San Diego, College Hill Press, 1984, pp 213–240.
8 Kelly JP, Khanna SM: Ultrastructural damage in cochleas used for studies of basilar membrane mechanics. Hear Res 1984;14:59–78.
9 Kelly JP, Khanna SM: Distribution of cochlear damage caused by the removal of the round window membrane. Hear Res 1984;16:109–126.
10 Koester CJ: Scanning mirror microscope with optical sectioning characteristics: Applications in ophthalmology. Appl Opt 1980;19:1749–1757.
11 Willemin JF, Dändliker R, Khanna SM: Heterodyne interferometer for submicroscopic vibration measurements in the inner ear. J Acoust Soc Am 1988;83:787–795.
12 International Team for Ear Research: Cellular vibration and motility in the organ of Corti. Acta Otolaryngol (Stockh) Suppl 1989.
13 Ulfendahl M, Flock Å, Khanna SM: Cochlear micromechanics – from isolated cells to the intact hearing organ. Proc 2nd Int Symp Cochlear Mechanics Otoacoustic Emissions, Rome 1989. Adv Audiol. Basel, Karger, 1990, vol 7, pp 27–34.
14 Khanna SM, Ulfendahl M, Flock Å: Mechanical response of the outer hair cell region of the isolated guinea pig cochlea in vitro; in Wilson JP, Kemp DT (eds): Cochlear Mechanisms – Structure, Function, and Models. New York, Plenum Press.
15 Kiang, NYS, Watanabe T, Thomas EC, Clark LF: Discharge patterns of single fibers in the cat's auditory nerve. Research Monograph Cambridge, MIT Press, 1965, vol 35.
16 Brownell WE, Bader CR, Bertrand D, de Ribaupierre Y: Evoked mechanical responses of isolated cochlear outer hair cells. Science 1985;227:194–196.
17 Zenner HP, Zimmermann U, Kepp U: High frequency motility of outer hair cells. Arch Otorhinolaryngol 1986;243:343–344.
18 Flock Å, Flock B, Ulfendahl M: Mechanisms of movement in outer hair cells and a possible structural basis. Arch Otorhinolaryngol 1986;243:83–90.

19 Ulfendahl M: Volume and length changes in outer hair cells of the guinea pig after potassium induced shortening. Arch Otorhinolaryngol 1988;245:237–243.

20 Slepecky N, Ulfendahl M, Flock Å: Effects of caffeine and tetracaine on outer hair cell shortening suggest intracellular calcium involvement. Hear Res 1988;32:11–22.

21 Canlon B, Brundin L, Flock Å: Acoustic stimulation causes tonotopic alterations in the length of isolated outer hair cells from guinea pig hearing organ. Proc Natl Acad Sci 1988;85:7033–7035.

22 Strelioff D, Flock Å: Stiffness of sensory-cell hair bundles in the isolated guinea pig cochlea. Hear Res 1984;15:19–28.

Shyam M. Khanna, PhD, Department of Physiology II, Karolinska Institutet, S–104 01 Stockholm (Sweden)

Grandori F, Cianfrone G, Kemp DT (eds): Cochlear Mechanisms and
Otoacoustic Emissions. Adv Audiol. Basel, Karger, 1990, vol 7, pp 27–34

Cochlear Micromechanics – From Isolated Cells to the Intact Hearing Organ[1]

Mats Ulfendahl[a], Åke Flock[a], Shyam M. Khanna[b]

[a] Department of Physiology, Karolinska Institutet, Stockholm, Sweden;
[b] Department of Otolaryngology, College of Physicians and Surgeons,
Columbia University, New York, N.Y., USA

Antibody labelling techniques have been used to identify and localize
actin at the electron-microscopic level in cochlear outer hair cells. Actin is
found to be present not only in the sensory hair region but also along the
wall of outer hair cells, between the plasma membrane and the fenestrated
cisternae, which are present here [1]. In order to investigate if outer hair
cells show some form of contractile response, isolated cells were subjected
to media that would induce contraction of muscle cells. In response to Ca^{2+}
and ATP, hair cells showed a significant shortening. These experiments
were performed on cells demembranated by a detergent. In intact cells,
electron microscopy showed that the plasma membrane is intimately
linked to subsurface cisternae by membrane-associated rodlets that remind
of the connection between the transverse tubule system and the sarcoplasmic reticulum in skeletal muscle fibers. The sarcoplasmic reticulum is
electrically excited by the muscle action potential to release Ca^{2+} which
triggers the contractile event. This event can be mimicked by depolarizing
the muscle cell with high concentrations of potassium salts. The same
treatment causes a shortening of the outer hair cells [2]. Caffeine is another
compound which has a shortening effect both on muscle cells and outer
hair cells [3]. Outer hair cells can also be induced to shorten or elongate in
response to electrical stimulation [4]. Another triggering mechanism of
motile events was recently demonstrated when it was shown that length
changes in isolated outer hair cells could be induced in response to a

[1] This research was supported by a program project grant NS 22334 from NIDCD,
the Swedish Medical Research Council 0246 and the Tysta Skolan and Söderberg Foun-
dations.

mechanical signal applied to the cell membrane [5]. Since this stimulus is more related to the physiological mechanical stimulation of the hair cells provided by the motion of the cochlear partition in response to sound, these experiments may indicate that each tone reaching the cochlea could produce motile events within the hearing organ. It is thus of importance to study the effects of motile events within the intact organ of Corti. To address this topic, we recently designed an in vitro preparation of the guinea pig temporal bone [6] to be used for measuring the cochlear vibration pattern with the laser heterodyne interferometer [7]. This preparation not only proved to be a valuable tool for studying the effects of chemical manipulation of outer hair cell motile mechanisms but has in addition given new and interesting information on cochlear micromechanics, especially with respect to the role of the outer hair cells. This paper presents the guinea pig temporal bone preparation and the first results from experiments where the hearing organ was subjected to caffeine, a compound known to elicit shortening of isolated cells.

Temporal Bone Preparation

Guinea pigs are decapitated, the left temporal bone is excised and attached to a metal tube using dental cement. The metal tube serves both as a mechanical support and a coupling between the sound stimulus system and the external auditory meatus (fig. 1). While keeping the temporal bone moist to prevent if from drying, the bulla is opened to expose the bony cochlea, the tympanic membrane and the malleus. After transferring the preparation to the experimental chamber which is filled with 200–400 ml of tissue culture medium, the thin bone overlying the most apical turns is carefully removed (fig. 2a) so that Reissner's membrane and the stria vascularis are left intact. During the experiment the cochlea and the middle ear are immersed in the fluid of the medium.

The dissection leaves the tympanic membrane, the middle ear ossicles and the oval window intact and these structures are used for transmitting the sound stimulus to the cochlea. The sound source is connected to the metal tube-auditory meatus with flexible rubber tubing (fig. 1).

The exposed cochlear turn is viewed using an optical sectioning microscope [8]. The preparation is viewed horizontally using a 20 × water immersion lens inserted in the chamber through a rubber gasket. A working distance of 4 mm gives sufficient mechanical access to view different parts of the cochlea and to observe the malleus. The metal tube and an aluminium rod connect the preparation to a precision orientation system. This system allows positioning and tilting of the preparation. The microscope uses incident light illumination and the optical sectioning technique ensures that the tissue can be observed at a specific depth without interference from tissue above or below the focal plane. The resulting image gives a detailed view of the reticular lamina where many of the cellular elements can be identified (fig. 2b).

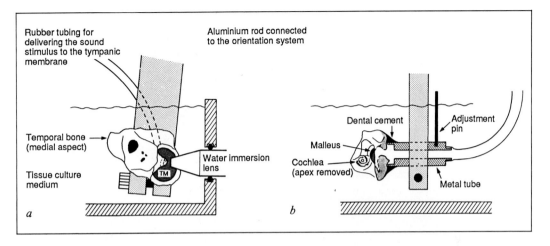

Fig. 1. Schematic illustration of the temporal bone preparation viewed from the medial side (a) and the anterior side (b).

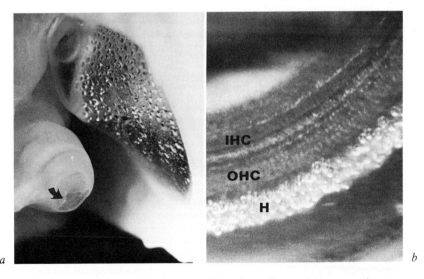

Fig. 2. a The opened bulla showing the cochlea, the malleus and the tympanic membrane. Parts of the cochlear wall lining scala vestibuli are removed to expose the organ of Corti (arrow). b The organ of Corti as seen in the optical sectioning microscope. The image is obtained through the intact Reissner's membrane and the regions of the inner and outer hair cells (IHC, OHC) and Hensen's cells (H) can be identified.

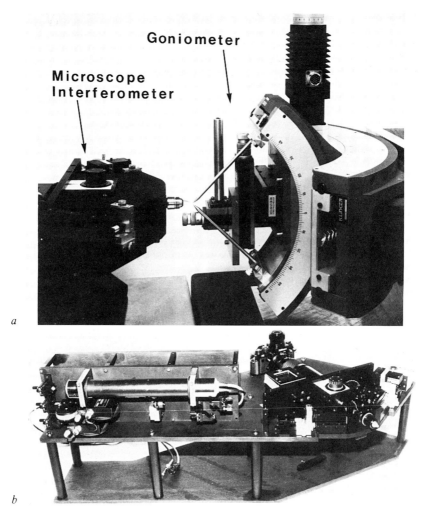

Fig. 3. A photograph showing the optical sectioning microscope, the laser hetero-dyne interferometer and the orientation (goniometer) system.

To measure the mechanical response of the cochlear partition a laser heterodyne interferometer is used. The interferometer is integrated with the optical sectioning micro-scope (fig. 3), so that the laser beam (helium-neon) is focused at the same layer of the tissue as observed in the microscope. By determining the Doppler shift due to the vibra-tion of the observed structure, the velocity amplitude is measured. The spot size of the laser beam is less than 5 μm in diameter and thus the velocity can be obtained at the single cell level, i.e. measurements can be made on a selected specific cell within the organ.

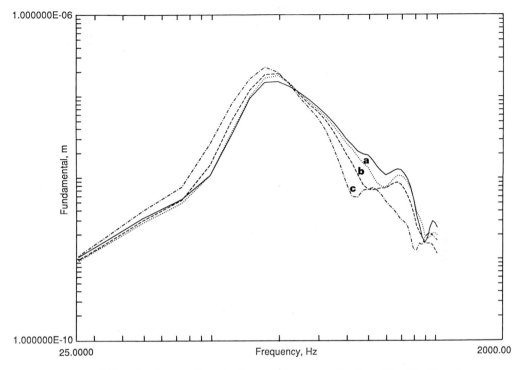

Fig. 4. The vibration amplitude before and after the application of 5 m*M* caffeine to the isolated temporal bone. When caffeine is applied the vibration amplitude increases. Solid line, before caffeine; a = 30 min, b = 60 min, and c = 90 min after the application of caffeine.

Despite the low reflectivity of living tissue, the interferometer is sensitive enough to allow measurements of the vibration amplitude below 1 Å.

Details of the temporal bone preparation, the optical sectioning microscope and the laser heterodyne interferometer are presented elsewhere [6–8].

Caffeine and Cochlear Mechanics

One of the substances used for studying the influence of motile behaviour of outer hair cells on the vibration pattern of the organ of Corti is caffeine. This compound has been used for many years in muscle physiology for its ability to induce the release of calcium ions from the sarcoplasmic reticulum and thus to induce muscle contraction [9–11]. Caffeine

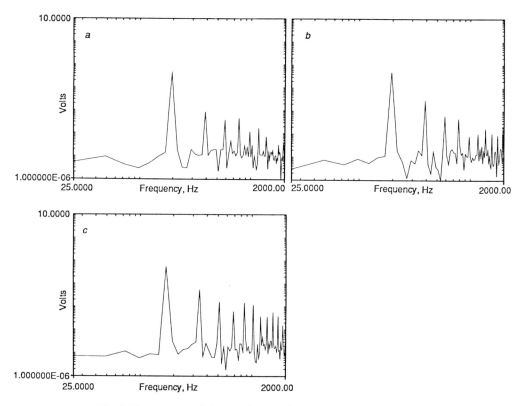

Fig. 5. The spectrum of the velocity waveform before (*a*) and 30 and 90 min after the application of caffeine (*b*, *c*). Note the increase in both the number and the amplitudes of the higher harmonics.

induces within minutes a sustained shortening of isolated outer hair cells, possibly by a similar mechanism [3].

To test the effects of outer hair cell shortening on cochlear microme-chanics, caffeine (5 m*M*) was applied to the isolated temporal bone. This resulted in profound changes in the vibration pattern of the organ of Corti. The vibration amplitude of the mechanical response at the fundamental frequency increased slightly in the presence of caffeine (fig. 4). This finding differs from all other experiments where the vibration amplitude tends to decline with time. It was also observed that the characteristic frequency first moved up and then down in frequency during the experiment. The

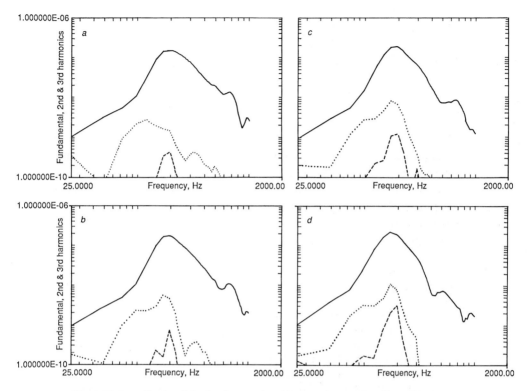

Fig. 6. The amplitude of the fundamental and harmonics as a function of frequency and time. Before and after application of caffeine. The fundamental is shown with a solid line, the second harmonic with a dotted line and the third harmonic with a dashed line. Before exposure (*a*), after 30 min (*b*), after 60 min (*c*) and after 90 min of exposure (*d*).

most interesting finding, however, was that the tuning of the mechanical response (the fundamental) became sharper in the presence of caffeine. This effect was first observed at about 30 min after the application of caffeine and continued throughout the experiment. At the same time, the content of higher harmonics increased (fig. 5). The parallel change in the shape of the mechanical response and in the increase of the amplitudes of the second and third harmonics is illustrated in figure 6.

The effects on the vibration amplitude and the tuning of the response resulting from the application of a drug known to affect outer hair cells imply a more active role for these cells than previously thought. In addition, these experiments are interesting in that, for the first time, they show the effect of activation of the outer hair cell motile mechanisms on the cochlear micromechanics and especially the tuning mechanism.

Conclusions

The above results lead to the conclusion that outer hair cells in the organ of Corti possess a calcium release and uptake mechanism which controls the contractile state of these cells. It is suggested that outer hair cells can exhibit motor activity, which is integrated in the electromechanical function of the organ of Corti. The results from the caffeine experiments suggest that the shortening of outer hair cells affects the sensitivity, tuning frequency, frequency selectivity and non-linearity in the cochlea.

References

1   Flock Å, Flock B, Ulfendahl M: Mechanisms of movement in outer hair cells and a possible structural basis. Arch Otorhinolaryngol 1986;243:83–90.
2   Ulfendahl M: Volume and length changes in outer hair cells of the guinea pig after potassium-induced shortening. Arch Otorhinolaryngol 1988;245:237–243.
3   Slepecky N, Ulfendahl M, Flock Å: Effects of caffeine and tetracaine on hair cell shortening suggest intracellular calcium involvement. Hear Res 1988;32:11–22.
4   Brownell WE, Bader CR, Bertrand D, de Ribaupierre Y: Evoked mechanical responses of isolated cochlear hair cells. Science 1985;227:194–195.
5   Canlon B, Brundin L, Flock Å: Acoustic stimulation causes tonotopic alterations in the length of isolated outer hair cells from the guinea pig hearing organ. Proc Natl Acad Sci, USA 1988;85:7033–7035.
6   Ulfendahl M, Flock Å, Khanna SM: A temporal bone preparation for the study of cochlear micromechanics at the cellular level. Hear Res. 1989;40:55–64.
7   Willemin JF, Dändliker R, Khanna SM: Heterodyne interferometer for submicroscopic measurements in the inner ear. J Acoust Soc Am 1988;83:787–795.
8   Koester CJ: Scanning mirror microscope with optical sectioning characteristics: Application in ophthalmology. Appl Opt 1980;19:1749–1757.
9   Axelsson J, Thesleff S: Activation of the contractile mechanism in striated muscle. Acta Physiol Scand 1958;44:55–66.
10   Bianchi CP: The effect of caffeine on radiocalcium movement in frog sartorius. J Gen Physiol 1961;44:845–858.
11   Weber A, Herz R: The relationship between caffeine contracture of intact muscle and the effect of caffeine on reticulum. J Gen Physiol 1968;52:750–759.

Dr. Mats Ulfendahl, Department of Physiology II, Karolinska Institutet, S–104 01 Stockholm (Sweden)

Grandori F, Cianfrone G, Kemp DT (eds): Cochlear Mechanisms and
Otoacoustic Emissions. Adv Audiol. Basel, Karger, 1990, vol 7, pp 35–41

Fast and Slow Motility of Outer Hair Cells in vitro and in situ

Hans Peter Zenner, Günter Reuter, Peter K. Plinkert, Alfred H. Gitter

HNO Hearing Research Laboratories, University of Tübingen, FRG

OHCs can move in the presence of depolarization [6] and electrical [1, 2], chemical [8, 9] or mechanical [22] stimuli.

Fast Outer Hair Cell Motility

Isolated outer hair cells (OHCs) have been reported to oscillate in the presence of electrical stimuli [1–3] up to at least 30 kHz. The underlying mechanism is unclear.

In situ experiments with cochleas from guinea pigs [4] showed that an electrically stimulated OHC within the organ of Corti (OC) is capable of longitudinal movements along its cylindrical cell body. At the same time, a radial displacement of the cuticular plate (CP) including the bundle of stereocilia could be observed. A limited number of adjacent OHCs moved synchronously with the stimulated OHC.

By means of photodiode techniques (fig. 1), electrically generated movements of the OHCs could be followed up to 10 kHz. The activity within the OC appeared to be restricted to the three rows of the OHCs. By mechanical coupling, OHC stimulated inner hair cells (IHCs) to radial movements.

Thus, OHCs are able to move actively in the OC during electrical stimulation (fig. 2). A shortening of the cell body is transformed into a radial shear movement of the stereociliary bundle via a tilting movement of the CP. The CP 'slides' laterally in the plane of the RL. At the same time, the RL is also moved in the direction of the basilar membrane (BM). The radial movement is transferred in part to the IHCs via 'RL coupling'.

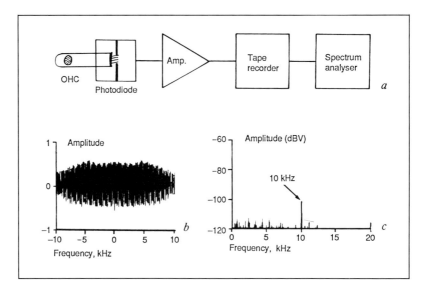

Fig. 1. High frequency motility of isolated OHCs. *a* OHC movements are measured by a double photodiode which is used as a linear position detector. *b* Signal (autocorrelation function) of an OHC, of which the apical pole is moving between the two photodiodes with 10,000 Hz. Stimulation is an extracellular a.c. field (1 μV/μm) along the OHC. *c* Frequency analysis (power density function) of the signal in b. From reference 2.

This is evidently reduced by a compliance between the inner row of OHCs and the tunnel of Corti (TC). Furthermore, these shear motions could be the basis of fluid and/or tectorial membrane (TM) coupling of OHCs and IHCs thus allowing a mechanical signal transfer from OHCs to IHCs. It is conceivable that the pillar cells and Deiter's cells, including their processes extending up to the RL, constitute a passive three-dimensional matrix which codetermines the complex movement sequence described for the OHCs.

If these fast movement sequences of the OHCs play a role under physiological conditions similar to the experimental active motile responses in situ, these might amplify the transverse vibratory movements of the BM. Coupling of both the radial and the transverse OHC movements to the BM can be explained by a torque [5]. The active radial component with displacements of the CP including the stereocilia might serve to amplify the mechanical signal by reinforcing the TM/fluid coupling to the stereocilia of

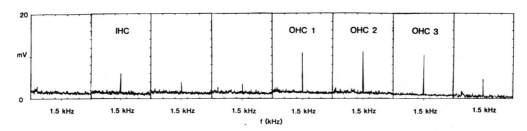

Fig. 2. Radial 1.5-kHz movements in the OC. A cochlear explant was exposed to an a.c. field (1–5 µV/µm) along (± 20°) the OHC in the OC. Radial movements were measured using the double photodiode shown in figure 1 as linear position detector at the locations indicated. The power density spectra show pronounced 1.5-kHz responses of the three rows of OHCs and a smaller response of the IHCs. All other spectra are from supporting cells, revealing minute movements.

the IHC. Results with isolated OHCs revealed an association of OHC shortening with a depolarization and of OHC elongation with re- and hyperpolarization [6]. Thus, a sound-induced deflection of OHC stereocilia towards Hensen's cells with subsequent OHC depolarization could initiate this amplification process near auditory threshold. A localized mechanical amplification process by a group of OHCs might contribute to otoacoustic emissions and to nonlinearity, negative damping and sharp tuning of the BM motions [7].

Biphasic Slow Motile Responses of Outer Hair Cells

A slow depolarization of isolated OHCs (e.g. in the presence of 25–125 mM K) is accompanied (fig. 3) by a slow biphasic motile response of the cell body and the cuticular plate (CP) displacing the stereociliary bundle [6]. Furthermore, a perilymphatic perfusion with elevated K^+ concentrations induced a shortening of the OHCs in vivo. Histological controls revealed a reduced distance between BM and reticular lamina (RL) [Fermin and Zenner, in preparation]. Thus the depolarization of OHCs in situ is accompanied by motile OHC events inducing a compression of the OC. Single-frame recordings using digital image subtractions allowed to differentiate two phases. During the first 0.5–5 s OHCs shorten and move their CP. A longer exposure to elevated K^+ concentrations results in an additional increase of the cell volume, which is accompanied by a further

Fig. 3. Longitudinal elongation and shortening of an isolated OHC. The unusually pronounced motile response was achieved by a slow depolarization of the OHC.

length change of cell. The experiments suggest two separate underlying mechanisms: an early phase and possibly voltage-dependent electromechanical process and a late-phase volume regulatory mechanisms of OHCs. Own FURA measurements revealed that a depolarization of hair cells is associated with an increase of the intracellular Ca^{2+} level. We were able to show biochemically that an increase of the intracellular calcium level activated the actomyosin skeleton of isolated OHCs [8, 9]. This resulted in slow contractions of the cell body and the CP with a velocity of 3–24 nm $\times$ ms^{-1} during a first phase of 50–200 ms. Longer stimulations induced an additional longer second phase of the motile response with time constants in the second to minute range. Results in the minute range were also obtained by others [10]. We suggest that only the first-phase motile responses play a physiological role, whereas the second phases represent interesting pathological conditions.

Furthermore we could show that the intracellular Ca^{2+} level of OHCs is also controlled by inositol-trisphosphate (IP_3) acting as a second messenger in hair cells [11]. Thin-layer chromatography revealed that OHCs contain a cascade of phosphoinositides, and thus the biochemical machinery to produce IP$_3$ in their outer cell membranes. An experimental increase of intracellular IP$_3$ in isolated OHCs induced motile responses during 50–200 ms. Thus IP$_3$ can act as a second messenger to elicit a motile response, the molecular basis of which is thought to be the rise of the intracellular Ca^{2+} level and the subsequent activation of actomyosin.

A direct involvement of IP$_3$ in the electromechanical process seems unlikely, as in other cell species the release of IP$_3$ is mostly a separate agonist-receptor-operated process.

Recently we found acetylcholine receptors (AChR) in the outer cell membrane of OHCs [12]. The presence of GABA receptors has been postulated [13, 14]. Monoclonal antibodies against internal and external epitopes of the AChR from torpedo allowed us a direct visualization of AChRs in whole-cell preparations from OHCs [12]. AChRs could be found exclusively in the basal and basolateral membrane parts of OHCs. Sometimes, they were arranged like a cup at the basal end of the sensory cell. Furthermore, the monoclonal antibodies showed patching of AChRs suggesting a possible lateral diffusion of AChRs in the membrane plane of OHCs. Moreover, labelling by the antibodies of AChRs was found on both sides of the outer cell membrane, indicating the presence of a pleiotropic, transmembrane-protein. Thus, our results suggest the presence of AChRs at the synaptic end of OHCs.

We hypothesize that via ACh efferent nerves activate these newly found AChRs. This could result in an agonist-receptor-operated release of IP$_3$ from the membrane-bound phosphoinositide PIP$_2$ into the cytoplasm, thus generating a Ca^{2+}-controlled response.

The first-phase slow longitudinal and shear movements are suggested to control baseline position and/or compliance of the OC in response to sound. They do it by pushing or pulling on BM and RL allowing a bias of the RL/BM location and thus of the operation point of the hair bundles. Thus, by Ca^{2+}-dependent actomyosin-driven movements of CP and cell body, OHCs could e.g. compress and decompress BM and hair bundle displacements during high sound pressure levels (SPLs). This mechanism is thought to contribute to adaptation processes (automatic gain control) or temporary threshold shifts (TTS) [8, 9, 15]. Furthermore it could set an operation point of the OC at high SPL reducing the general sensitivity of a broad band of IHCs below the actual SPL. This can be accompanied by fast motile OHC responses superimposing the slow mechanical event. In contrast to the slow movements, the fast mechanical OHC responses are sharply focussed thus allowing frequency selectivity even during background noise at elevated SPLs. Thus at high SPL slow and fast OHC motility might cooperate to allow signal filtering and to improve signal to noise ratio. LePage [16] speculated that the spatial dependence of a long-term control of the BM bias might define the tuning curve as a function of tone level. The size of long-term baseline movements may be very much larger than that of the travelling wave vibration at threshold sound levels. This behavior is voltage-, IP_3- and AChR-controlled and affected by the efferent system.

Moreover, the 'muscle-like' [17] OHCs could follow the sound stimulus cycle by cycle for frequencies less than approximately 1 kHz. A similar mechanical phase-locking has been proposed for low-frequency coding, making possible the velocity response and phase-locking in neural period histograms [16]. For sound frequencies above 1 kHz on the basis of models [7] and our own biochemical results, we have suggested that the actomyosin skeleton might significantly help amplify the travelling wave by intermittent stimulations of a resonance system [8, 9, 18]. This cannot be excluded to be a possible cellular basis for otoacoustic emissions.

Bidirectional Transduction in Outer Hair Cells

All motile mechanisms of OHCs must lead to an alteration of the sound-induced displacement of the stereocilia and must thus influence the auditory process directly by closing a feedback loop [3, 19, 20]. Thus an electromechanical transduction process has been suggested [3, 19, 20]. Together with the mechanoelectrical transduction, this gives rise to a bi-directional transduction cycle in OHCs. The efferent fibres of the auditory nerve, which have numerous endings on the OHCs [21], may well interfere with this feedback system [3].

References

1   Brownell WE, Bader CR, Bertrand D, de Ribaupierre Y: Evoked mechanical responses of isolated cochlear outer hair cells. Science 1985;227:194–196.
2   Zenner HP, Zimmermann U, Gitter AH: Fast motility of isolated mammalian auditory sensory cells. Biochem Biophys Res Commun 1987;49:304–308.

3   Zenner HP, Arnold W, Gitter AH: Outer hair cells as fast and slow cochlear amplifiers. Acta Otolaryngol (Stockh) 1988;105:457–462.

4   Reuter G, Zenner HP: Imaging of active motile responses in organ of Corti. Hear Res. In press, 1989.

5   Zwislocki JJ: Mechanical properties of the tectorial membrane in situ. Acta Otolaryngol (Stockh) 1988;105:450–456.

6   Zenner HP, Zimmermann U, Schmitt U: Reversible contraction of isolated mammalian cochlear hair cells. Hear Res 1985;18:127–133.

7   Zwicker E: A model describing non-linearities in hearing by active processes with saturation at 40 dB. Biol Cybern 1979;35:243–250.

8   Zenner HP: Motile responses in outer hair cells. Hear Res 1986;22:83–90.

9   Zenner HP: Motility of outer hair cells as an active, actin-mediated process. Acta Otolaryngol (Stockh) 1988;105:39–44.

10   Flock Å, Flock B, Ulfendahl M: Mechanisms of movement in outer hair cells and a possible structural basis. Arch Otorhinolaryngol 1986;243:83–90.

11   Schacht J, Zenner HP: Evidence that phosphoinositides mediate motility in cochlear outer hair cells. Hear Res 1988;31:155–160.

12   Plinkert PK, Gitter AH, Zimmermann U, Kirchner T, Tzartos S, Zenner HP: Visualization and functional testing of acetylcholine receptors in cochlear outer hair cells. Hear Res. In press, 1989.

13   Klinke R: Die Verarbeitung von Schallreizen im Innenohr. HNO 1987;35:139–144.

14   Pujol R, Lenoir M: The four types of synapses in the organ of Corti; in Altschuler RA, Hoffmann DW, Bobbin RP (eds), Neurobiology of Hearing: The Cochlea. New York, Raven Press, 1986.

15   Zenner HP: Aktive Bewegungen von Haarzellen: Ein neuer Mechanismus beim Hörvorgang. HNO 1986;34:133–138.

16   LePage EL: Functional role of the olivocochlear bundle: A motor unit control system in the mammalian cochlea. Hear Res. In press, 1989.

17   Zenner HP: Cytoskeletal and muscle-like elements in cochlear hair cells. Arch Otorhinolaryngol 1980;230:82–92.

18   Zenner HP, Zimmermann R, Gitter AH: Active movements of the cuticular plate induce sensory hair motion in mammalian auditory hair cells. Hear Res. In press.

19   Kim DO: Active and nonlinear cochlear biomechanics and the role of outer hair cell subsystems in the mammalian auditory system. Hear Res 1986;22:105–114.

20   Zenner HP, Gitter AH: Active motility and reverse transduction in outer hair cells; in Löbe LP, Lotz P (eds): Cochlea Research. Halle, University Press, 1987.

21   Spoendlin H: Innervation densities of the cochlea. Acta Otolaryngol (Stockh) 1972;73:235.

22   Canlon B, Brundin L, Flock Å: Acoustic stimulation causes tonotopic alterations in the length of isolated outer hair cells from the guinea pig hearing organ. Proc Natl Acad Sci USA 1988;85:7033–7035.

Dr. H.P. Zenner, Department of Otolaryngology, University of Tübingen, D–7400 Tübingen (FRG)

Grandori F, Cianfrone G, Kemp DT (eds): Cochlear Mechanisms and
Otoacoustic Emissions. Adv Audiol. Basel, Karger, 1990, vol 7, pp 42–46

Physiopathology of Sensory Hair Cells: In vivo and in vitro Studies on Aminoglycoside Uptake and Toxicity

Jean-Marie Aran[1]

Laboratoire d'Audiologie Expérimentale, Unité de Recherche INSERM 229 et
Université de Bordeaux II, Hôpital Pellegrin, Bordeaux, France

Biomechanical interactions in sensory hair cells of the cochlea are at
the basis of mechanoneural transduction. These interactions are particularly important for outer hair cells, which receive and generate vibrations
[1]. Such mechanisms can be directly affected by many physiopathological
processes, such as acoustic overstimulation or the toxicity of aminoglycoside molecules which, for instance, have been shown to interfere with
transduction by blocking the K^+ channels at the top of the stereocilia in
lower vertebrates [2].

In this paper we intend to review and discuss, at the cellular level, our
current data obtained in vivo and in vitro on the toxic mechanisms of
aminoglycosides and their relation with the function of the hair cells.

Acoustic Potentiation, Cellular Uptake and Development of Ototoxicity

Many studies have demonstrated a synergistic effect of association of
acoustic stimulation with aminoglycoside treatment [3]; however, all of
them have used sound levels and aminoglycoside doses close to their toxic
levels. In recent experiments on guinea pigs with chronically implanted

[1] The author wishes to thank Yves Cazals, Didier Dulon, Jerôme Dupont, Hakim
Hiel and Kathleen Horner for their valuable discussions during the elaboration of the
manuscript. The study was supported in part by the 'Conseil Régional d'Aquitaine'.

electrodes in the round window or the auditory cortex, performed in our laboratory [4], it appeared that the specific ototoxicity observed after the administration of a combination of gentamicin (GM) with a loop diuretic such as ethacrynic acid (EA) under normal sound environment could be significantly delayed and sometimes totally reduced when the animal was kept in the quiet. The treatment consisted of 150 mg/kg GM i.m. followed 1.5 h later by 30 mg/kg EA i.v. or i.c. Sound exposure in the unprotected animals was completely atraumatic: they were exposed only to 70 dB pe SPL click presented at 10/s for 25.6 s every 5 min over 3–4 h to monitor the fast changes in compound action potential (CAP) amplitude after the EA injection.

Thereafter the guinea pigs were kept in the animal quarters (noise level around 60 dB SPL) and tested occasionally for frequency threshold curves over several days. A second group of guinea pigs was kept in the quiet and just exposed to the same click stimulations only during the first 10 min in order to check the effectiveness of the EA injection. Thereafter they were kept in a double-wall sound proof room (IAC 1202A) and the frequency threshold curve was measured immediately before sacrifice. Our data are in contrast with the other studies where sound exposure and drug treatment were both just below toxic levels [3]. These observations clearly indicate that normal activation of the hair cells definitely facilitates the development of ototoxicity, in other words silence affords a certain protection against aminoglycoside ototoxicity.

Recent immunohistological and autoradiographic studies [4–6] have shown that aminoglycosides administered either as a chronic treatment or 'acutely' after potentiation with a loop diuretic, as described above, penetrate specifically into the hair cells, the radial and spiral distribution of the aminoglycosides corresponding to their electrophysiologically and morphologically well-documented regional specificity [4].

Moreover, in our experiments, GM could be clearly identified inside the outer hair cells of the entire first turn in cochleae which had not shown any functional change, as determined by the very sensitive measure of thresholds of auditory cortex or eighth-nerve evoked potentials as a function of frequency. This was true in guinea pigs after treatment under different conditions. First, following combined GM/EA treatment and thereafter in the quiet [4]. Secondly, following chronic treatment with daily injection of 60 mg/kg GM, in normal sound environment and sacrificed between the 6th and 12th day [Hiel et al., in preparation]. This later treatment, when administered over 14 days, leads to significant threshold elevations and hair cell loss. Thus it appears that GM uptake by hair cells can occur before or even without threshold changes. This could explain the sometimes delayed development of ototoxicity described clinically [7].

These various observations suggest that cellular uptake is potentiated by acoustic activation of the cells and that toxicity is an intracellular process which might be facilitated by functional depolarization of the cell. An important issue in understanding the toxic mechanisms is the route of entry of aminoglycoside antibiotics into the hair cells. Much evidence suggests that they could enter from the endolymph, through the apical surface of the cell:

The time course of the development of functional and morphological changes seems to parallel (follow) that of the entry of the drug into the endolymph: although the drug is found very early in the perilymph, ototoxic signs develop only after the drug has penetrated into the endolymph. GM entry into the endolymph and ototoxicity are delayed during chronic treatment with GM alone while both are immediate when GM is administered in combination with EA [8-10]. The kinetics of GM in the perilymph (entry and clearance) are rapid, parallel that in serum and are unchanged under EA [10, 11]; however, in this later case, EA could also facilitate the entry of GM directly into the hair cell from the perilymph (cortilymph).

At the earliest stage, GM is found precisely and only in the area of the basal body, below the apical surface of the cell deprived of a cuticular plate [4], a place thought to be suited for metabolic exchanges with the endolymph.

In vitro studies on isolated outer hair cells have recently demonstrated that even several hours after aminoglycosides have been added to the culture medium (1) there is no modification of the viability of outer hair cells, nor of their motility in response to K^+ depolarization or electrical stimulation [12, 13]; (2) there is no penetration of the molecule inside the hair cells, and labelling is observed only extracellularly in the efferent synaptic area [12, 13] but (3) voltage-sensitive calcium channels are blocked [13, Dulon et al., in preparation].

The lack of penetration and of effects on viability and motility in vitro is in agreement with the absence of penetration and of functional changes following a single intramuscular injection of the aminoglycoside in vivo, although it is present in the perilymph within 2–3 h after the injection [10]. In this condition transient functional changes, if any, could remain undetected due to the fast kinetics of the aminoglycoside in the perilymph. Either the aminoglycoside molecule is penetrating into the hair cell through the lateral cell wall membrane very slowly, or it does not enter the hair cell from the perilymph. The functional effect on calcium channels, which has also been indirectly observed in in vivo experiments [14], would be nontoxic and could be reversible as in the case of the neuromuscular blockage, which has been shown to be reversed by calcium chloride [15] and seems to be related to acetylcholine release [16]. Moreover the neuromuscular blockage potency of aminoglycosides [17] appears to be inversely proportional to their ototoxic potency [8]: the degree of neuromuscular blocking potency of netilmicin = sisomicin > neomicin > gentamicin > tobramycin. Thus any effect due to exposure of the hair cell to the aminoglycoside from perilymph might be of the neuromuscular blockage type and related to the efferent control of the outer hair cells. The effect of exposure from endolymph would be basically different, leading to penetration into the cell at its apical end and development of intracellular toxicity.

The potentiation by acoustic stimulation of both uptake and toxicity demonstrates the basic interference of aminoglycosides with transduction processes both at the channel site(s) and intracellularly.

References

1   Davis H: an active process in cochlear mechanics. Hear Res 1983;9:79–90.
2   Howard J, Roberts M, Hudspeth AJ: Mechanoelectrical transduction by hair cells. Annu Rev Biophys Chem 1988;17:99–124.
3   Collins PWP: Synergistic interactions of gentamicin and pure tones causing cellular hair cell loss in pigmented guinea pigs. Hearing Res 1988;36:249–260.
4   Hayashida T, Hiel H, Dulon D, Erre JP, Guilhaume A, Aran J-M: Dynamic changes following combined treatment with gentamicin and ethacrynic acid with and without acoustic stimulation. Acta Otolaryngol (Stockh) 1989 (in press).
5   Hayashida T, Nomura Y, Iwamori M, Nagai Y, Kurata T: Distribution of gentamicin by immunofluorescence in the guinea pig inner ear. Arch Otorhinolaryngol 1985;242:257–264.
6   Veldman JE, Meeuwsen F, van Dijk M, Key Q, Huizing EH: Progress in temporal bone histopathology. II. Immuno-technology applied to the temporal bone. Acta Otolaryngol (Stockh) 1985;suppl 423:29–35.
7   Glorig A: Clinical manifestations of ototoxicity and noise. Adv Otorhinolaryngol 1973;20:2–13.
8   Aran J-M, Erre JP, Guilhaume A, Aurousseau C: The comparative ototoxicities of gentamicin, tobramycin and dibekacin in the guinea pig. A functional and morphological cochlear and vestibular study. Acta Otolaryngol (Stockh) 1982;suppl 390: 1–30.

9   Aran J-M, Darrouzet J: Observation of click evoked compound VIII nerve responses before, during and over several months after kanamycin treatment in the guinea pig. Acta Otolaryngol (Stockh) 1975;79:24–32.

10   Tran Ba Huy P, Manuel C, Meulemans A, Sterkers O, Amiel C: Pharmacokinetics of gentamicin in perilymph and endolymph of the rat as determined by radioimmunoassay. J Infect Dis 1981;143:476–486.

11   Tran Ba Huy P, Manuel C, Meulemans A, Sterkers O, Wassef M, Amiel C: Ethacrynic acid facilitates gentamicin entry into endolymph of the rat. Hear Res 1983; 11:191–202.

12   Dulon D: Ototoxicité des antibiotiques aminoglycosidiques: cinétique d'incorporation tissulaire in vivo et approche cellulaire in vitro; thèse Bordeaux 1987.

13   Dulon D, Zajic G, Schacht J: Acute effects of aminoglycosides on calcium flux, motility and viability of isolated outer hair cells; in Lim DJ (ed): Abstr 12th Midwinter Res Meet of ARO, 1989.

14   Bernard PA, Bourret C: Perilymphatic calcium and VIIIth nerve action potentials during gentamicin bolus i.v. injections. A preliminary study. Acta Otolaryngol (Stockh) 1987;103:400–403.

15   Paradelis AG, Crassaris LG, Karachalios DN, Triantaphyllidis CJ: Aminoglycoside antibiotics: Interaction with trimethaphan at the neuromuscular junctions. Drugs Exp Clin Res 1987;13:233–236.

16   Paradelis AG, Triantaphyllidis C, Giala MM: Neuromuscular blocking activity of aminoglycoside antibiotics. Methods Find Exp Clin Pharmacol 1980;2:45–51.

17   Rutten JM, Booij LH, Rutten CL; Crul JF: The comparative neuromuscular blocking effects of some aminoglycoside antibiotics. Acta Anaesth Belg 1980;31:293–306.

Jean-Marie Aran, MD, Laboratoire d'Audiologie Expérimentale,
Unité de Recherche INSERM No. 229 et Université de Bordeaux II,
Hôpital Pellegrin, F–33076 Bordeaux (France)

Grandori F, Cianfrone G, Kemp DT (eds): Cochlear Mechanisms and
Otoacoustic Emissions. Adv Audiol. Basel, Karger, 1990, vol 7, pp 47–56

Otoacoustic Emissions in Frogs

J.P. Wilson[a], *R.J. Baker*[a], *M.L. Whitehead*[b]

[a] Department of Communication and Neuroscience, University of Keele,
Staffordshire, UK; [b] Department of Otorhinolaryngology and Communicative
Sciences, Baylor College of Medicine, Houston, Tex., USA

It is now generally accepted that otoacoustic emissions (OAEs) are a byproduct of an active process whose primary function is to sharpen up the inherent poor tuning of the basilar membrane. This view, first expressed by Gold [1], is supported by evidence from OAEs themselves [2] and from recent direct measurements of basilar membrane motion [3, 4] and in particular, its extreme physiological vulnerability. Several mechanisms have been proposed to produce the required motility including the piezoelectric effect [5], volumetric changes in hair cells [6], length changes in hair cells [7], tilting of the cuticular plate [8], tilting of the stereocilia [9]. Some of the features of the OAE appear to suggest the outer hair cell as its source. It is therefore of considerable interest to look at other hearing organs where the structure and interaction of different elements may be different. In particular there is now considerable evidence that certain hearing organs rely on electrical rather than mechanical tuning [10, 11]. Supporting this dichotomy is the finding that some hearing organs show strong temperature dependence of tuning [12, 13], whereas others such as mammals [14, 15] show remarkably little sensitivity to temperature. It is therefore of considerable interest to study any differences in the behavior of frequency tuning mechanisms in different species. It is intended here to review some of the data on OAEs in frogs.

OAEs in frogs were first reported by Palmer and Wilson [16] in *Rana esculenta*. Evoked emissions were demonstrated for both click stimulation

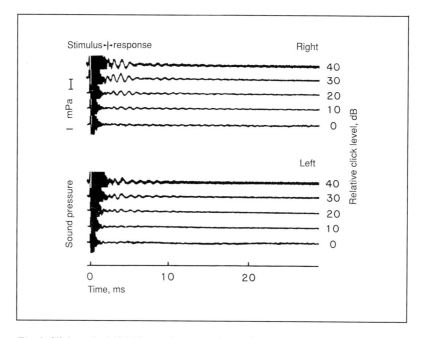

Fig. 1. Click-evoked (OAE) sound pressure in the frog, *R. esculenta,* as a function of time. The response gain is kept constant for a series of 10-dB increments in the stimulus. Owing to the short latency, the response is not separated from the stimulus (blackened area) at the higher levels [from ref. 16].

(fig. 1) and for continuous tonal stimulation (fig. 2), and spontaneous emissions (SOAEs) were found in the 1-kHz region in 7 out of 12 frogs.

The latency to peak response is about 4–4.5 ms at 1 kHz which is appreciably shorter than the 15 ms found for man for a similar frequency [17]. The frequency range for a continuously stimulated emission clearly extends from 0.8 kHz to at least 2 kHz, possibly 2.8 kHz. This would appear to implicate both the amphibian and basilar papillae. The frog also appeared to be an efficient generator of OAEs as at the lower levels (5 dB SPL) the re-emission could be only about 6 dB below the stimulus. The maximum re-emitted level of about 10 dB SPL is slightly lower than that of man. The phase delays calculated from the slopes of the straight segments of amplitude minima (at 0.7, 1, 1.3, 1.7, 2.1 kHz) are 3.3 ms and 2.5 ms at 1 and 1.7 kHz, respectively. The former is not inconsistent with the group delay (but see later).

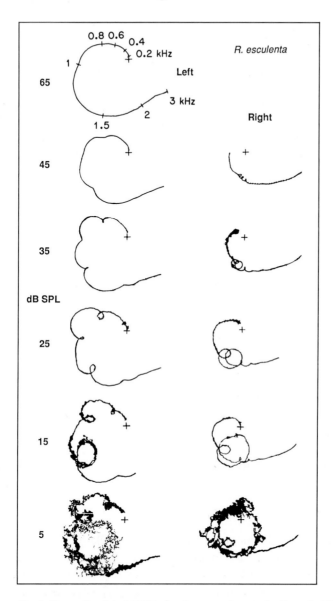

Fig. 2. Tone-stimulated OAEs for the same frog as in figure 1. The Nyquist plots represent the locus of vectors (origin +) with the frequency positions along the curve indicated in the upper left, highest stimulus level plot. As the stimulus frequency changes along the curves, the delayed response moves in and out of phase with the stimulus giving loops and cusps. At the lower sound levels the re-emitted sound vector becomes relatively larger on the normalised plots leading to larger cusps and loops in the curves [from ref. 16].

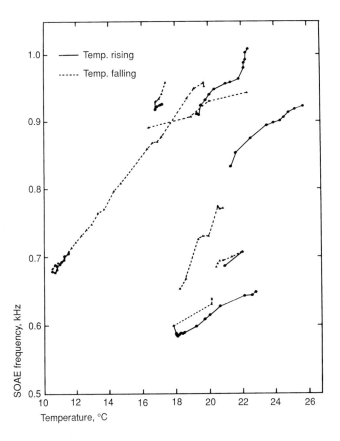

Fig. 3. Temperature dependence of the frequency of SOAEs in *R. temporaria.* Lines connect data points in time sequence with continuous lines rising and dashed lines falling temperature [from ref. 18].

The apparent involvement of the basilar papilla implied by the response at 2 kHz or above led Whitehead et al. [18] to investigate the effect of temperature on OAEs to see whether there was correspondence with the neural findings of Moffat and Capranica [19] where the amphibian papilla showed a strong temperature dependence whereas the basilar papilla did not. Ten *Rana temporaria* were anaesthetised with MS222 (0.4 mg/kg i.p.). Sound was measured with a B & K 4179 1-inch low noise microphone and analysed with lock-in amplifiers. Both spontaneous and tone-stimulated OAEs were investigated as body temperature

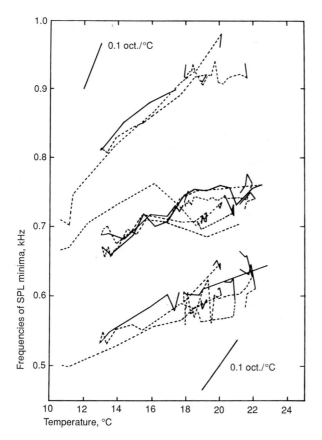

Fig. 4. Temperature dependence of the frequency of SPL minima in *R. temporaria.* Continuous curves rising, dashed curves falling temperature. Note that the slopes are less than the 0.1 oct/ °C found in caiman neurones [13] [from ref. 18].

was manipulated by an electrical heater or surrounding ice cubes and monitored by a miniature thermocouple in the mouth. In the case of stimulated emissions the frequency measured was that producing a sound pressure minimum due to destructive interference between the stimulus and the emission.

In every case a strong positive dependence of frequency upon temperature was found (fig. 3, 4). Rather surprising, however, a considerable variation of temperature coefficient was observed ranging from 9 to 48 Hz/ °C (0.01–0.07 oct/ °C) for SOAEs and from 5 to 27 Hz/ °C (0.01–0.03 oct/ °C) for SPL minima. The variability greatly exceeded the errors of measurement and the repeatability of any one measurement. Furthermore the SOAE temperature dependence is approximately twice that for SPL minima. As it is believed that these arise through similar mechanisms, it is difficult to see how this could arise. One of the main aims of this study was, however, frustrated by a failure to find OAEs at higher frequencies where it would be certain that they were arising from the basilar papilla. Over the range of frequencies available (600–1,000 Hz at 20 °C) there was no tendency for higher frequencies to have a lower temperature coefficient. These changes are, however, much greater than could be expected for a mechanical system where tuning frequency depends on mass and stiffness, and are indicative of electrochemical tuning.

More recently, these temperature effects have been confirmed in *R. esculenta* by van Dijk and Wit [20] who also showed that the increase in sound level of an SOAE with temperature was consistent with an Arrhenius plot of 7 eV/molecule, which they interpret as the chemical energy source for an active mechanism. They also showed that the amplitude distributions of an SOAE were characteristic of an active oscillator rather than filtered noise.

Wit et al. [21] were able to measure an electrical correlate of SOAEs with an electrode inserted in a small hole on the cranial side of the otic capsule. Interestingly, an electrical correlate of the distortion product, f_2-f_1, could also be detected. This lay in the 0.7- to 0.9-kHz range for an SOAE of 0.9 (frequency shifted) to 1.1 kHz and an external tone of 1.8 kHz. This, however, is above the range of neural distortion product frequencies (0.2–0.55 kHz) found by Capranica and Moffat [22] and is therefore puzzling.

Baker et al. [23] looked specifically for evidence of any form of nonlinear behavior in OAEs in frogs. A wide range of primary frequencies and distortion product frequencies were investigated in *R. temporaria, Rana pipiens,* and *R. esculenta* but in no case was a distortion product (f_2-f_1, $2f_1-f_2$, $2f_1$, $2f_2$, $2f_2-f_1$) detected that was above instrumental distortion level. At 80 dB SPL this was at – 70 dB and for 60 dB SPL it was below the noise floor at – 80 to – 85 dB. Specifically, no trace of distortion could be detected under the conditions where it had been measured neurally by Capranica and Moffat [22] and electrically by Wit et al. [21]. The disagreement with the former findings, however, may simply correspond with the failure to find any kind of OAE below 600 Hz in frogs, possibly due to poor middle-ear coupling. It is also possible that OAEs in frogs arise from certain features of the structure of the anuran ear and therefore only occur in certain frequency regions (i.e. excluding the frequency region 200–550 Hz where neural evidence of distortion exists). This view would appear to be supported by the great uniformity of OAE properties between animals of the same species and the apparent existence of three separate regions of emission. The former finding is in contradistinction to mammals (particularly man) where the waveform for click emissions, or its corresponding spectrum for tonal stimulation, is specific to each ear. This has been interpreted as being due to either one or more mechanical impedance discontinuities on the basilar membrane [2, 24] or one or more irregularities in the place/frequency map [25]. In frogs the hair cells are embedded in the limbic wall and not supported on a basilar membrane [26] so the second hypothesis would appear more probable. Furthermore, the existence of two separate hearing organs and of two distinct neural regions for the amphibian papilla [22] would appear to provide in-built 'mapping discontinuities'. In fact, there appears to be a tendency for the optimal frequencies of emission to occur in three groups at about 600, 725 and 950 Hz at 20 °C (see fig. 3, 4). This suggestion should, however, be treated with caution as there is no reason why the maximum gain of the active process should occur near either the SOAE frequency or a sound pressure minimum; it is the phase shift characteristics which contribute most strongly to the location of these frequencies. The other area of doubt is whether, in the frog, the frequency separation between minima should be considered as due to an inherent time delay (as seems to be the case in mammals) or is simply due to the existence of three separate discontinuities. The data are not yet sufficiently accurately defined to be decisive on this point.

The other types of nonlinear behavior which are exhibited are a saturating input/output function where the response increases by about 0.5 dB/1 dB input, suppression of either an SOAE (fig. 5a, b) or an evoked OAE by another tone, and the slight but consistent shift in SOAE frequency produced by a 'suppressing' tone. This manifests itself as a shift of the SOAE away from the 'suppressing' frequency (except that the neutral point is actually about 100 Hz below the SOAE frequency, see fig. 5c, d). This kind of frequency shift also occurs in man [27] but to a lesser extent. In both cases the threshold for a frequency shift lies at or below the threshold for suppression. As yet no satisfactory explanation for it has been proposed. In general terms, the nonlinear input/output function and the suppression tuning curves appear very similar in frog and man although the tuning is somewhat broader in frog.

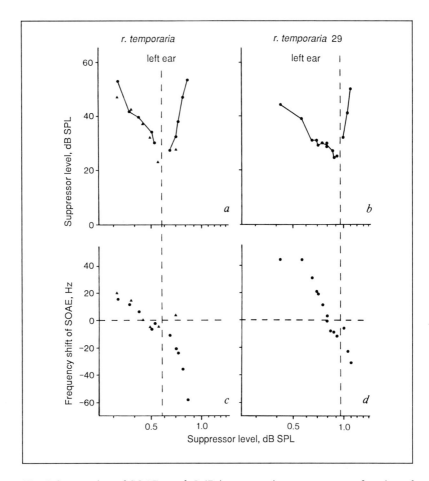

Fig. 5. Suppression of SOAEs. *a, b* 5-dB isosuppression contours as a function of suppressor frequency for SOAEs located at the frequencies indicated by the vertical dashed lines. *c, d* Shift of SOAE frequency as a function of suppressor frequency at the 5-dB suppression level. Note that the neutral point (no frequency shift) lies about 100 Hz below the SOAE frequency [from ref. 23].

Some of the minor differences described above are scarcely surprising in view of the very different structures in frog and mammal. On the other hand, many of the common features would appear to imply a common origin. This may be some common motile property of hair cells which transcends the question of whether they have electrical tuning properties or not.

References

1   Gold T: Hearing II. The physical basis of the action of the cochlea. Proc R Soc B 1948;135:492–498.
2   Kemp DT: Stimulated acoustic emissions from within the human auditory system. J Acoust Soc Am 1978;64:1386–1391.
3   Khanna SM, Leonard DGB: Basilar membrane tuning in the cat cochlea. Science 1982;215:305–306.
4   Sellick PM, Patuzzi, R, Johnstone BM: Measurements of basilar membrane motion in the guinea pig using Mössbauer techniques. J Acoust Soc Am 1982;72:131–141.
5   Naftalin L: Some new proposals regarding acoustic transmission and transduction. Cold Spring Harbor Symp Quant Biol 1965;30:169–180.
6   Wilson JP: Model for cochlear echoes and tinnitus based on an observed electrical correlate. Hear Res 1980;2:527–532.
7   Brownell WE: OHC motility and cochlear frequency selectivity; in Moore BCJ, Patterson RD (eds): Auditory Frequency Selectivity. New York, Plenum Press, 1986, pp 109–122.
8   Zenner HP: Motile responses in outer hair cells. Hear Res 1986;22:83–90.
9   Crawford AC, Fettiplace R: The mechanical properties of ciliary bundles of turtle cochlear hair cells. J Physiol 1985;364:359–380.
10 Crawford AC, Fettiplace R: An electrical tuning mechanism in turtle cochlear hair cells. J Physiol 1981;212:377–412.
11 Pitchford S, Ashmore JF: An electrical resonance in hair cells of the amphibian papilla of the frog, Rana temporaria. Hear Res 1987;27:75–83.
12 Eatock RA, Manley GA: Temperature effects on single auditory nerve fibre responses. J Acoust Soc Am 1976;60:s80.
13 Smolders J, Klinke R: Effect of temperature on the primary auditory fibres of the spectacled caiman, Caiman crocodilus. J Comp Physiol 1984;155:19–30.
14 Gummer AW, Klinke R: Influence of temperature on tuning of primary like units in the guinea-pig cochlear nucleus. Hear Res 1983;12:367–380.
15 Wilson JP: The influence of temperature on frequency-tuning mechanisms; in Allen JB, Hall JL, Hubbard A, Neely ST, Tubis A (eds): Peripheral Auditory Mechanisms. Berlin, Springer, 1985, pp 229–236.
16 Palmer AR, Wilson JP: Spontaneous and evoked acoustic emissions in the frog Rana esculenta. J Physiol 1981;324:66p.
17 Wilson JP: Evidence for a cochlear origin for acoustic re-emissions, threshold fine structure and tonal tinnitus. Hear Res 1980;2:233–252.
18 Whitehead ML, Wilson JP, Baker RJ: The effects of temperature on otoacoustic emission tuning properties; in Moore BCJ, Patterson RD (eds): Auditory Frequency Selectivity. New York, Plenum Press, 1986, pp 39–48.
19 Moffat AJM, Capranica RR: Effects of temperature on the response properties of auditory nerve fibres in the American toad (Bufo americanus). J Acoust Soc Am 1976;60:s80.
20 van Dijk P, Wit HP: Temperature dependence of frog spontaneous otoacoustic emissions. J Acoust Soc Am 1987;82:2147–2150.

21   Wit JP, van Dijk P, Segenhout JM: An electrical correlate of spontaneous otoacous-
     tic emissions in a frog: A preliminary report; in Wilson JP, Kemp DT (eds): Cochlear
     Mechanisms: Structure, Function and Models. New York, Plenum Press 1989, pp
     341–347.
22   Capranica RR, Moffat AJM: Nonlinear properties of the peripheral auditory system
     of anurans; in Popper AN, Fay RR (eds): Comparative Studies of Hearing in Verte-
     brates. New York, Springer, 1980, pp 139–165.
23   Baker RJ, Wilson JP, Whitehead ML: Otoacoustic evidence for nonlinear behaviour
     in frogs' hearing: Suppression but no distortion products; in Wilson JP, Kemp DT
     (eds): Cochlear Mechanisms: Structure, Function and Models. New York, Plenum
     Press 1989, pp 349–358.
24   Kim DO: Cochlear mechanics: Implications of electrophysiological and acoustical
     observations. Hear Res 1980;2:297–317.
25   Sutton GJ, Wilson JP: Modelling cochlear echoes: The influence of irregularities in
     frequency mapping on summed cochlear activity; in DeBoer E, Viergever M (eds):
     Mechanics of Hearing. Delft, Delft University Press, 1983, pp 83–90.
26   Geisler CD, van Bergeijk WA, Frischkopf LS: The inner ear of the bullfrog. J Mor-
     phol 1964;114:43–58.
27   Wilson JP, Sutton GJ: Acoustic correlates of tonal tinnitus; in Evered D, Lawrenson
     G (eds): Tinnitus. Ciba Foundation Symposium. London, Pitman Medical, 1981,
     No 85, pp 82–107.

J.P. Wilson, PhD, Department of Communication and Neuroscience,
University of Keele, Staffordshire, ST5 5BG (UK)

Grandori F, Cianfrone G, Kemp DT (eds): Cochlear Mechanisms and
Otoacoustic Emissions. Adv Audiol. Basel, Karger, 1990, vol 7, pp 57–62

Some Properties of the Cubic Distortion Tone Emission in the Guinea Pig

B.M. Johnstone, B. Gleich, N. Mavadat, D. McAlpine, S. Kapadia

Department of Physiology, University of Western Australia, Nedlands, Australia

The basic properties of the otoacoustic distortion product emission, $2f_1$-f_2, in the guinea pig have been described by Brown [1]. We have carried out further studies in the guinea pig in order to understand the origins of the high-level and low-level emissions and to quantify further the relationship between the low-level emission and the hearing threshold.

Some of the work was performed with a very simplified system using a lock-in amplifier. This system had a noise floor of about 10 dB SPL. Other work was performed with a more complex low-noise system using an HP 3561A analyzer. This system had a noise floor of -15 dB SPL.

A typical input-output function is shown in figure 1. The cubic distortion tone (2,800 Hz) was produced by equal intensities of the primary tones of 4,150 and 5,500 Hz. Also shown in figure 1 is the effect of loud sound (approximately 100 dB SPL) exposure for 5 min at four different frequencies of 10, 6, 4 and 3 kHz. No effect is seen for the two higher frequencies, a shift to the right is evident for the 4-kHz exposure and a further shift for the 3-kHz exposure. These exposures produced approximately 20 dB of threshold shift (TTS) at and above a half octave higher than the exposure frequency. We conclude that the $2f_1$-f_2 is generated at the local primary points and that TTSs more basal are not reflected in either the low-level or high-level distortion product. Also shown in figure 1 is the postmortem $2f_1$-f_2. It appears to be very similar to the presumed high-level distortion tone. If we give extensive TTS, we are left with a high-level distortion product growing approximately in a cubic

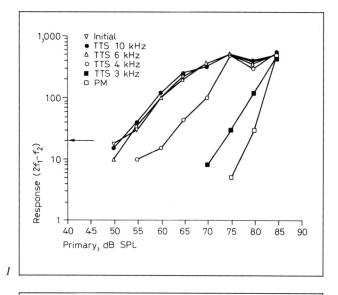

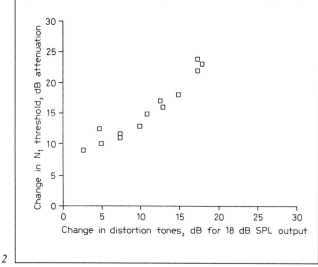

Fig. 1. Input-output functions for a 2,800 Hz distortion tone generated from primaries at 4,150 and 5,500 Hz. The vertical axis is in arbitrary units and the arrow indicates
18 dB SPL. There were 4 loud tone exposures in the order of 10, 6, 4 and 3. PM is the
10-min postmortem function.

Fig. 2. Correlation between changes in N_1 threshold after TTS and changes in the
primary levels needed to keep the distortion tones at 18 dB SPL (frequencies as in figure 1).

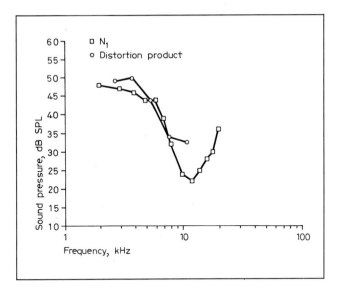

Fig. 3. Comparison of N_1 threshold with the – 10 dB SPL contour, i.e. the sound pressure of the primaries necessary to produce a distortion tone at – 10 dB SPL. The ratio of the primaries was 1:1.2. This animal had a small threshold loss in the low and midfrequencies.

manner, which is not changed for many minutes postmortem but slowly decays over an hour or so.

Using loud sounds it is possible to create variable threshold losses of about equal amounts at the primary tone frequencies. These threshold losses have been correlated with corresponding changes in the distortion product, measured as the changes in primary tones to maintain a 15-dB SPL distortion tone. The results shown in figure 2 are evidence of an excellent correlation over a 20-dB range with signs of a wall (or fence) at 30 dB loss which corresponds to the distortion tone reaching the postmortem value. More recent work using our improved system and a distortion tone level of – 10 dB SPL shows a 30- to 35-dB range. It appears that a distortion tone of – 10 dB SPL corresponds to the CAP (N_1) threshold. Figures 3 and 4 show the CAP audiogram from two guinea pigs, one (fig. 3) with a slight midfrequency loss and one (fig. 4) with a large notch centered at 8 kHz. Both of these animals' losses were adventitious with unknown etiology. Also shown is the – 10-dB SPL distortion tone contour. Note the close

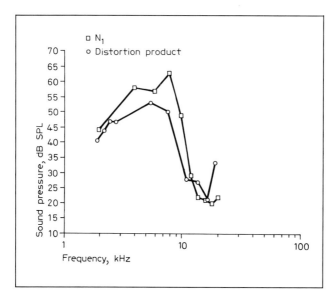

Fig. 4. Same as figure 3 but here the animal has a 10- to 30-dB loss between 6 and 12 kHz.

correspondence up to 14 kHz. The higher frequencies show a large spread, possibly due to difficulties in matching the two primary levels.

Measuring the distortion tone at low frequencies shows some interesting results. The input-output function for primaries of 255 and 300 Hz (fig. 5) show almost no difference between the alive and postmortem animal. This behavior becomes less evident around 500 Hz and by 750 Hz a normal behavior returns.

The postmortem distortion tone is subject to TTS as shown in figure 6. The amount of reduction (15 dB) is not much less than would be expected from the same exposure to a normal animal (about 20–25 dB).

We conclude that the cubic distortion product is a good measure of outer hair cell function and by inference of basilar membrane vibration, if measured at low levels. There are two sources of the distortion tone. One is due to the active process in the outer hair cells and is proportional to threshold over a 30-dB range. The other is a high level product and we hypothesize that it is produced by the nonlinear mechanical gating mechanism as described by Howard and Hudspeth [2]. The effect of TTS on both mechanisms would be to close the gates, thus eliminating their effects.

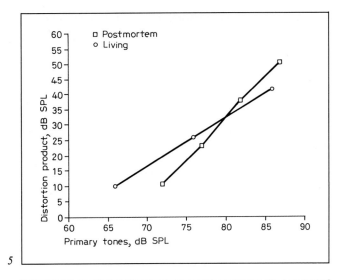

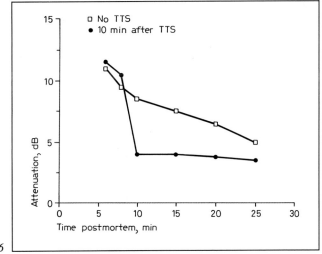

Fig. 5. Distortion tone production at low frequencies (f_1 = 255 Hz, f_2 = 300 Hz) compared with a 10-min postmortem production.

Fig. 6. The postmortem distortion tones for 18 dB SPL ($2f_1$-f_2) are plotted as a function of time. The vertical axis is arbitrary, with the zero attenuation corresponding to approximately 90 dB SPL. The frequencies are as for figure 1. Both animals were normal and were killed by an overdose of anesthetic. Zero time postmortem corresponds to the cessation of heart beat and total N_1 loss. In one animal (no TTS) the distortion tone was simply measured; in the other animal, a loud sound (3 kHz, 105 dB SPL for 3 min) was presented 10 min postmortem.

The very small amount of low-level distortion tone at low frequencies and its absence of change with death suggests that the active process is not acting at these frequencies in the guinea pig.

References

1   Brown AM: Acoustic distortion from rodent ears: A comparison of responses from rats, guinea pigs and gerbils. Hear Res 1987;31:25–38.
2   Howard J, Hudspeth AJ: Compliance of the hair bundle associated with gating of mechano-electrical transduction channels in the bullfrog's saccular hair cell. Neuron 1988;1:189–199.

B.M. Johnstone, MD, Department of Physiology,
University of Western Australia, Nedlands 6009 (Australia)

Grandori F, Cianfrone G, Kemp DT (eds): Cochlear Mechanisms and
Otoacoustic Emissions. Adv Audiol. Basel, Karger, 1990, vol 7, pp 63–76

Otoacoustic Emissions in Research of Inner Ear Signal Processing[1]

Eberhard Zwicker

Institute of Electroacoustics, Technische Universität München, FRG

Neurophysiological data of signal processing in hearing stem almost
exclusively from animals. Only in very rare cases, as during operations, are
there possibilities to collect data from humans. Most psychoacoustic data
stem from human subjects. Therefore, most decisions regarding the kind of
disease in hearing-impaired subjects are based on psychoacoustic data,
although objective results, for example impedance measurements, seem to
be more acceptable. From this point of view, otoacoustic emissions
(OAEs), which are based on physical measurements, with the subjects
being totally passive, seem to be an attractive tool to obtain information
about signal processing within the human inner ear. Such measurements
are especially effective for normal hearing, because OAEs appear almost
exclusively in subjects who have a hearing loss smaller than about 20 dB.
When OAEs are available, there is a large variety of strategies and methods
to determine information about inner-ear signal processing.

The Four Kinds of Otoacoustic Emissions

There are four kinds of OAEs, which can be measured by closing up
the outer ear canal with a probe that contains not only a very sensitive
microphone but also a small transmitter to produce sound [1]. The microphone picks up the sound pressure in the closed cavity containing signals
that are not produced through the transmitter but appear as an additional
effect.

[1] Supported by the 'Deutsche Forschungsgemeinschaft' (SFB 204, 'Gehör').

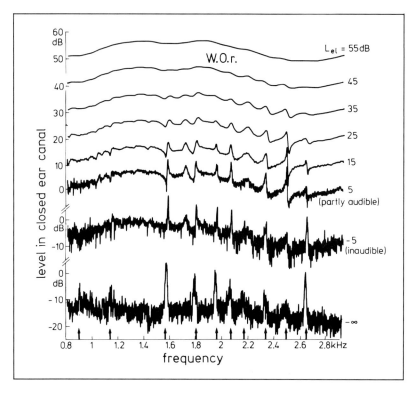

Fig. 1. The bottom trace shows the frequency-selective (Δf = 10 Hz) sound pressure level measured in the closed ear canal without any stimulus (L_{el} = $-\infty$). The arrows indicate SOAEs. The upper traces show the frequency-selective levels as responses to frequency-sweeping stimuli of given electrical level, L_{el}, at the transmitter of the probe. The resulting ripple in the responses indicates SEOAE.

Spontaneous Otoacoustic Emissions

Spontaneous OAEs (SOAEs) are the most spectacular emissions, because they are produced without any acoustic stimulation through the transmitter. With the subject as quiet as possible and using a low-noise microphone and a small analysing bandwidth of only a few hertz it is possible to get responses as low as -25 or even -30 dB SPL [2]. The frequency analysis of the sound pressure measured in the closed cavity without any stimulation reveals SOAEs in about 70% of normal-hearing human subjects. Such emissions are indicated as narrow spectral compo-

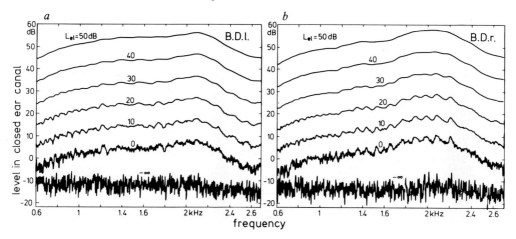

Fig. 2. The same as figure 1 for left and right ear of another subject who shows no pronounced SOAEs (see lowest trace).

nents of different level and frequency typical for each individual ear. The example outlined as the bottom trace in figure 1 shows a more unusual case of 10 SOAEs found in the frequency range between 0.8 and 2.8 kHz.

Simultaneously Evoked Otoacoustic Emissions

Simultaneously evoked OAEs (SEOAEs) are produced when a sinusoidal tone sweeping in frequency is transferred to the transmitter. The sound pressure level measured in the closed ear canal as a function of the evoking frequency with constant electrical input level, L_{el}, shows a relatively smooth and flat response at high levels (fig. 1, upper most trace with L_{el} = 55 dB). It is determined by the frequency response of the transmitter and the microphone in the closed cavity. Reducing the input level, however, produces responses in many subjects that show increasing additional ripple with decreasing level [3, 4]. Sometimes this ripple is visible up to levels that correspond to values near 40 or 50 dB above threshold. An example with a remarkable ripple in the frequency responses is shown in figure 1 for input levels between −5 and +45 dB. The ripple is strongly correlated to the frequencies of the SOAEs. However, subjects with no or very small SOAEs also show this frequency ripple very often. Figure 2 outlines SEOAEs of such a subject for the left and the right ear, which show very similar effects. The level dependence of the ripple is used as an indication

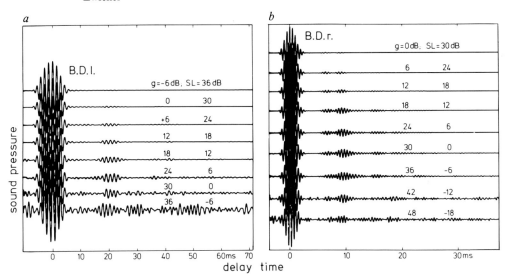

Fig. 3. DEOAE, i.e. sound pressure-time function in response to a sequence of short tone bursts, measured in the closed ear canal of the same subject as in figure 2. *a, b* show responses to 700-Hz and to 2.1-kHz tone bursts, respectively. Parameter is the sensation level (SL). Note that the gain (g) of the amplifier was reduced in the same manner as sensation level was increased, so that the evoker shows always the same amplitude. Note also the different time scales (abscissae) in *a* and *b*.

that there are SEOAEs, which appear in about 96% of normal hearing subjects [5].

Such emissions appear not only as level or amplitude responses but also as phase responses that correspond to each other typically in such a way that phase is increasing at frequencies for which the amplitude or the level shows a minimum. If the frequency and phase response of the system picked up at high levels is subtracted from what is measured at low levels, then a vector diagram of the SEOAE can be drawn [4]. It shows a very periodical structure running clockwise around the origin with a period of a frequency difference, Δf, that increases with frequency.

Delayed Evoked Otoacoustic Emissions

Delayed evoked OAEs (DEOAEs) are produced by short stimuli. The time function of the reponse is measured using a synchronous averaging process. The repetition rate of the tone burst or the click ranges between 10 and 50 Hz, corresponding to periods between 100 and 20 ms [6]. Within

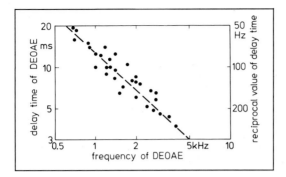

Fig. 4. Delay time of DEOAEs (left ordinate scale) as a function of the frequency of DEOAEs. The reciprocal value of the delay time is also indicated (right ordinate scale) and corresponds to the frequency difference between neighbouring maxima or minima of SEOAEs.

that time, DEOAEs reach their maximum. Figure 3 shows examples of relatively large emissions produced in the left and the right ear of the same subject, for which SEOAEs are indicated in figure 2. The emissions are evoked by Gaussian shaped 0.7- or 2.1-kHz tone burst sequences, the sensation level of which is a parameter in the different panels. At low levels, the emissions remain constant if – as indicated in figure 3 – the gain is reduced in the same way as the sensation level is increased. Taking this into account, figure 3 indicates that the emissions grow in the same manner as the evoking burst for low sensation levels. For sensation levels above about 15 dB, however, the emissions show decrease equivalent to the decrease of the gain, i.e. they remain constant in amplitude [7].

As for SEOAEs, a strong dependence of DEOAEs on the level of the evoking tone bursts can be observed. Such a dependence is a typical characteristic of evoked DEOAEs and is often used as their indication. DEOAEs are found in about 97% of normal hearing subjects [5]. The delay time of these emissions is dependent on their frequency as indicated in figure 4. While the left ordinate scale indicates the delay time, the right ordinate scale shows the reciprocal of this time difference, which leads to a frequency difference, directly related to that of neighboring maxima or minima of SEOAEs [8].

This frequency difference corresponds to a critical band rate distance of 0.4 Bark and can also be found when the probability of the frequency distance of neighboring SOAEs is transferred into the critical band rate

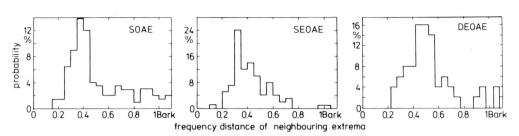

Fig 5. Probabilities of the frequency difference between neighboring extrema of SOAEs, SEOAEs, and DEOAEs; the latter are based on spectral analysis of the DEOAE time functions.

distance [9]. This probability is shown in figure 5 for all three kinds of emissions. It shows a strong peak in the neighborhood of 0.4 Bark, that seems to be an important characteristic value related to signal processing of the inner ear at low levels and determines also in many subjects the fine structure of threshold in quiet [9–12].

Distortion Product Otoacoustic Emissions

Distortion product OAEs (DPOAEs) can be measured when not only one but two stimulating tones are transferred to the closed cavity. In this case, the microphone picks up not only the two primaries but also additional tones, at frequencies that are typical for nonlinear distortion products [13]. Figure 6b shows an example. The level of the spectral components is given as a function of frequency. Equivalent measurements using a dummy coupler (fig. 6a) instead of a real ear indicate the high linearity of the system that is required to measure these relatively small effects. Using such systems it may be possible to use DPOAEs to determine cochlear characteristics of hearing-impaired subjects, because these emissions appear at high levels.

SOAEs indicate very clearly that our inner ear uses active feedback and a nonlinearity, which limits the amplitude of the internal oscillation to small inaudible values. The level dependence of the three kinds of emissions discussed first as well as the existence of DPOAEs also indicate that the effect of activity of our inner ear's signal processing behaves in a strongly nonlinear manner. These effects are hints for the strategies used by our inner ear. Such strategies have been simulated in models [14–16] which describe inner ear signal processing.

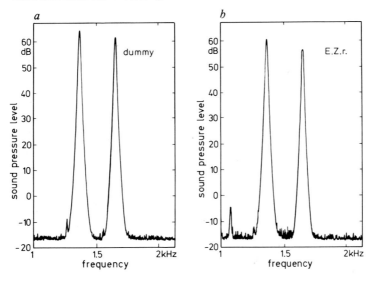

Fig. 6. Spectral analysis of a two-tone complex obtained in a dummy coupler *(a)* and in a subject's closed ear canal *(b)* causing an additional component at $2f_1-f_2$ that indicates a DPOAE.

Frequency Selectivity Measured by Otoacoustic Emissions

SOAEs can be reduced in level by adding a suppressor tone. Using a criterion such as the reduction of the SOAE level by 3 or 6 dB, a suppression tuning curve can be measured. It indicates the sound pressure level of the suppressor tone needed to reduce the emission by a fixed amount as a function of its frequency. Figure 7a shows such a curve for a 3-dB reduction of the SOAE marked by the circled cross. SEOAEs can also be reduced by an additional suppressor tone. Averaged data of such measurements are outlined in figure 7b. Both curves (fig. 7a, b) show similar dependence and are shaped in a form that is well known from neurophysiological and psychoacoustical tuning curves [17]. The shift of the dip of the tuning curve towards higher frequencies in relation to the frequency of the emission (indicated by the vertical dashed line in fig. 7b) is typical [5, 8]. Such effects are not only shown in neurophysiological tuning curves but also in psychoacoustical tuning curves. Other typical characteristics are the flat tail towards low frequencies and the steep rise towards higher frequencies, which often is interrupted by a pronounced second dip. Because OAEs are

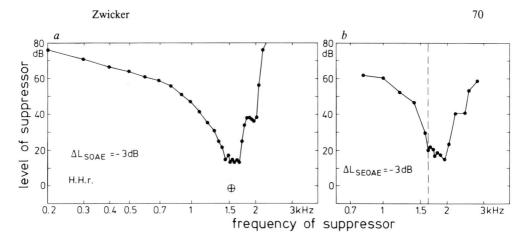

Fig. 7. Suppression tuning curves, i.e. level of a suppressor needed to reduce an SOAE *(a)* or an SEOAE *(b)* for a certain level (ΔL = – 3dB) as a function of the suppressor frequency. The circled cross in *a* indicates the SOAE, the vertical dashed line in *b* characterizes the frequency of the suppressed SEOAE.

assumed to be produced within the inner ear, it is obvious from these data that the frequency selectivity produced in the signal processing of the inner ear is almost exclusively responsible for the frequency selectivity measured in neurophysiological or even in psychoacoustical tuning curves. This means that OAEs are a perfect tool to measure the frequency resolution of our hearing systems.

Temporal Effects of Otoacoustic Emissions

Only two temporal effects may be described to concentrate within this review. These are the suppression period patterns, on the one hand, and the poststimulus time dependence on the other hand. Suppression period patterns are produced by DEOAEs when a periodical low-frequency suppressor is presented at the same time [18, 19]. Its frequency, when sinusoidal, or its fundamental frequency, when nonsinusoidal, has to be the same as the repetition rate of the short evoker signal leading to the DEOAEs. The psychoacoustically measured masked threshold of the evoking tone burst is also influenced by the masker's (suppressor's) time function. This influence is very strongly related to the influence of the masker/suppressor time function on the appearance of DEOAEs. Figure 8

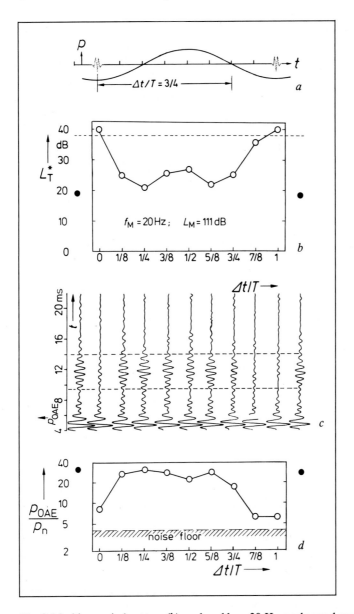

Fig. 8. Masking-period pattern *(b)* produced by a 20-Hz masker and measured using a 1,350-Hz tone burst (dotted) as indicated in *a*. Suppression period pattern *(d)* elaborated from the sound pressure-time functions *(c)* of DEOAEs evoked by the same test signal as indicated in *a* with a constant sensation level of 20 dB as marked by the dashed line in *b*. The data outside the frame are produced without masker or suppressor.

shows a typical example using a 20-Hz masker and 1.4-kHz tone bursts as test sound or evoker. It demonstrates together with data produced using nonsinusoidal maskers [7, 20] that the second derivative of the sound pressure time function of the masker is responsible for the effect of masking as well as for the effect of suppression of delayed emissions. This means that simultaneous masking is already installed within the inner ear's nonlinear signal processing.

A strong nonsimultaneous effect can be produced through the poststimulus dependence of either threshold in quiet or delayed emissions. When a high-level low-frequency stimulus appears for a period of 10 min, threshold in quiet, after switching off this stimulus, shows a strong dependence of poststimulus time. When DEOAEs are measured after the presentation of the same stimulus, the level of these emissions follows a temporal course that shows a time dependence almost exactly mirror-shaped of that of threshold in quiet. This means that such poststimulus effects are already produced within the inner ear also. However, these temporal courses have periods in the regions of minutes, indicating that the source of these effects may be related to the metabolism of the inner ear.

The Model and the Effect of Recruitment

Without going into detail, our models [14, 16] may be summarized by the fact that there is a passive linear system which is characterized besides the mass of the fluid of the inner ear by the elasticity of the basilar membrane and by its surrounding mass. Both last-mentioned values vary as a function of the location between oval window and helicotrema. In addition to that, there are nonlinear feedback loops, which are assumed to be realized through the outer hair cells' activity. The inner hair cells are assumed to play an important part only in transferring information about the displacement of basilar membrane towards higher centers through neural activity. The activity of the outer hair cells can be described as a kind of amplification with nonlinear saturating characteristic which acts through mechanical-electrical-mechanical transmission. The feedback loops, which also show lateral spread of activity, operate very close to the point of oscillation. However, the level of oscillations – if they occur as spontaneous emissions [15] – is limited to low levels through the nonlinearity of the amplifiers. In this way, a very high sensitivity as well as a distinct frequency selectivity are achieved. Additionally, the levels at the charac-

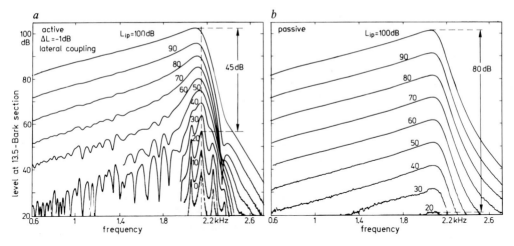

Fig. 9. Frequency selectivity of the model, i.e. level at a certain section (13.5 Bark corresponding to a characteristic frequency of 2,150 Hz) as a function of frequency for constant input level, L_{ip}. *a* Enhanced sensitivity and selectivity at lower input levels. *b* Linear behavior of the passive model.

teristic sections are strongly enhanced for low input levels up to 35 or even 45 dB in such a way that the high sensitivity appears only at low levels while the amplification in the feedback loop is reduced strongly at high levels and thereby almost negligible. The frequency response measured at a certain section of the analog model with the input level as parameter is indicated in figure 9a. It shows very clearly that low levels are enhanced so that a level decrement (ordinate) of only 45 dB appears at the characteristic frequency although the input level (parameter) decreases by as much as 80 dB. This holds only for the region of high sensitivity and not in the regions of the tail towards low frequencies where no compression takes place. Such a characteristic is produced in an active inner ear including nonlinear feedback. When, however, the outer hair cells are destroyed or their activities strongly reduced, then the inner ear does not act with enhancement but just linearly (see fig. 9b). This means that the relatively broadly tuned frequency response achieved at high levels is also achieved at lower and very low levels. When the system reacts passively, it is much less sensitive in relation to the active inner ear, as illustrated in figure 10a. If the healthy active inner-ear characteristic is used as a reference value, then the consequence of destroyed outer hair cells, i.e. switched off feedback loops, becomes obvious (see fig. 10b). The excitation level at the basilar mem-

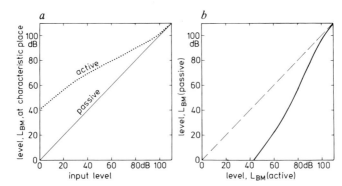

Fig. 10. a Relation between the level produced for CF at a certain section as a function of the input level as picked up from the data given in figure 9 for active and passive models. *b* Consequences of outer hair cell loss leading to a reduced sensitivity at lower levels comparable to the effect of recruitment.

brane available for active inner hair cells is shown as the abscissa, whereas the level of excitation produced by the same input level for conditions of destroyed outer hair cells is shown as the ordinate. At low levels, excitation is much smaller in the destroyed condition, whereas at high levels both the normal condition and the destroyed condition reach about the same values, an effect known as recruitment.

Conclusion

OAEs are an effective tool for research in inner ear signal processing. The models built so far show best agreement with measured data when nonlinear active feedback loops are installed that are assumed to be realized by the outer hair cells. Comparison of OAE data with model data indicates that not only the frequency selectivity but also simultaneous masking as well as nonlinear distortion products and temporary threshold shifts are created within the cochlea. Nonsimultaneous masking appears not to be installed in the inner ear and therefore is postulated to be post-synaptic [20, 22–24]. The effect of recruitment, however, seems to be created also within the cochlea through the fact that the outer hair cells are damaged or destroyed and therefore, cannot act as that part, which is responsible for enhancement of the excitation at low levels. This means that the very strongly increased sensitivity at low levels does not appear and, therefore, the human system is very insensitive for low input levels in such cases. At higher levels, however, the feedback loop is switched off automatically in normal conditions by the nonlinearity. Therefore, at higher levels, signal processing in inner ears with damaged outer hair cells appears almost equal to that for healthy inner ears.

References

1 Kemp DT: Stimulated acoustic emissions from the human auditory system. J Acoust Soc Am 1978;64:1386–1391.

2 Dallmayr C: Spontane otoakustische Emissionen, Statistik und Reaktion auf akustische Störtöne. Acustica 1985;59:67–75.

3 Kemp DT, Chum R: Properties of the generator of stimulated acoustic emissions. Hear Res 1980;2:213–232.

4 Zwicker E, Schloth E: Interrelation of different otoacoustic emissions. J Acoust Soc Am 1984;75:1148–1154.

5 Dallmayr C: Stationary and dynamic properties of simultaneous evoked otoacoustic emissions (SEOAE). Acustica 1987;63:243–255.

6 Wilson JP: Recording of the Kemp echo and tinnitus from the ear canal without averaging. J Physiol 1980;298:8–9.

7 Zwicker E: Delayed evoked oto-acoustic emissions and their suppression by Gaussian shaped pressure impulses. Hear Res 1983;11:359–371.

8 Zwicker E: The inner ear, a sound processing and a sound emitting system. J Acoust Soc Jpn 1988;9(E):59–74.

9 Zwicker E: Otoacoustic emissions and cochlear travelling waves; in Wilson JP, Kemp DT (eds): Cochlear Mechanics. New York, Plenum Press, 1989, pp 359–366.

10 Schloth E: Relation between spectral composition of spontaneous otoacoustic emissions and fine structure of threshold in quiet. Acustica 1983;53:250–256.

11 Long GR, Tubis A, Jones K: Changes in spontaneous and evoked otoacoustic emissions and the corresponding psychoacoustic threshold microstructure induced by aspirin consumption; in Allen JB, Hall JL, Hubbard AE, Neely ST, Tubis A (eds): Peripheral Auditory Mechanisms. Berlin, Springer, 1986, pp 213–218.

12 Wilson JP: Evidence for a cochlear origin for acoustic re-emissions, threshold fine structure and tonal tinnitus. Hear Res 1980;2:233–252.

13 Harris FP, Stagner BB, Martin GK, Lonsburgg-Martin BL: Effects of frequency separation of primary tones on the amplitude of acoustic distortion products. J Acoust Soc Am 1987;82(suppl 1):117.

14 Zwicker E: A model describing nonlinearities in hearing by active processes with saturation at 40 dB. Biol Cybern 1979;35:243–250.

15 Zwicker E: Oto-acoustic emissions in a nonlinear cochlear hardware model with feedback. J Acoust Soc Am 1985;80:153–162.

16 Lumer G: Computer model of cochlear preprocessing (steady state condition). I. Basis and results for one sinusoidal input signal. Acustica 1987;62:282–290.

17 Zwicker E: On a psychoacoustical equivalent of tuning curves; in Zwicker E, Terhardt E (eds): Facts and Models in Hearing. Berlin, Springer, 1974, pp 132–141.
18 Zwicker E: Masking-period patterns and cochlear acoustical responses. Hear Res 1981;4:195–202.
19 Zwicker E, Manley G: Acoustical responses and suppression-period patterns in guinea pigs. Hear Res 1981;4:43–52.
20 Zwicker E, Scherer A: Correlation between time functions of sound pressure, masking and OAE-suppression. J Acoust Soc Am 1987;81:1043–1049.
21 Zwicker E: A hardware cochlear nonlinear preprocessing model with active feedback. J Acoust Soc Am 1985;80:146–153.
22 Scherer A: Beschreibung der simultanen Verdeckung mit Effekten aus Mithörschwellen- und Suppressionsmustern. Fortschr Akustik, DAGA'87 Bad Honnef, DPG, 1987, pp 569–572.
23 Zwicker E: On peripheral processing in human hearing; in Klinke R, Hartmann R (eds): Hearing-Physiological Bases and Psychophysics. Berlin, Springer, 1983, pp 104–110.
24 Scherer A: Erklärung der spektralen Verdeckung mit Hilfe von Mithörschwellen- und Suppressionsmustern. Acustica. In press, 1989.

Dr. Eberhard Zwicker, Institute of Electroacoustics,
Technische Universität München, Arcisstrasse 21, D–8000 München 2 (FRG)

Grandori F, Cianfrone G, Kemp DT (eds): Cochlear Mechanisms and
Otoacoustic Emissions. Adv Audiol. Basel, Karger, 1990, vol 7, pp 77–98

Otoacoustic Emission Analysis and Interpretation for Clinical Purposes[1]

David T. Kemp, Siobhan Ryan, Peter Bray

Functional Analysis Laboratory, Audiology Department,
Institute of Laryngology and Otology, UCMSM, London, UK

A click-evoked otoacoustic test instrument was first proposed as an audiometric infant screening test by Kemp [1] in 1979. Figure 1 illustrates the basic technique. The proposal was based on work with normal and hearing-impaired adults [2]. Johnsen and Elberling [3] reported good results using the method on newborns. In other studies, mainly under laboratory conditions with adult subjects, it was confirmed that delayed evoked otoacoustic emissions (OAEs) were present in most normal human ears, and absent in most cases of cochlear impairment. The physiological significance of OAEs and their relation to cochlear mechanics is described elsewhere in this book. For the purpose of this practically oriented paper it is sufficient to note that registration of a wide-band click-evoked emission is highly indicative of a normal cochlea. The technique can confirm the integrity of a cochlear mechanism proven to be essential for a normal threshold, and to be extremely vulnerable to trauma and pathology. The reliable and precise clinical measurement and interpretation of otoacoustic emissions is therefore important in audiology, for screening, investigative, and research work.

From the outset otoacoustic emissions promised to be a quick objective test of cochlear function without electrodes. However, it has not proved easy to re-engineer the laboratory demonstrations of OAEs into a robust and reliable screening or diagnostic device. There were important

[1] This work was supported by the Medical Research Council, and by the Royal National TNE Hospital (BDHA) London. We also thank Otodynamics Ltd. for technical assistance with the production of the ILO88 apparatus.

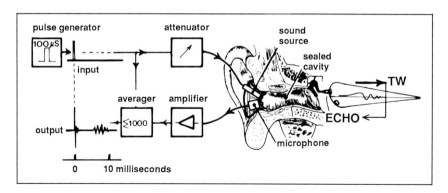

Fig. 1. The original basic experimental otoacoustic emission detector system. In the computer based system used to collect data for this paper the detector was enhanced by multiplexed stimulus level and coordinated subaveraging averaging, and a variety of analyses performed on the data.

technical problems to overcome such as the design of the probe, how to deal with acoustic artifacts, what stimuli should be used, and what parameters should be measured. In early practical trials of OAEs with clinical applications in mind, the above problems were addressed in an wide variety of different ways, or not addressed at all. The results were often confusing and disappointing.

The choice of otoacoustic techniques and parameters is wide. In 1986 Kemp et al. [4] presented an analysis of the possibilities for performing 'acoustic cochleography' using OAEs. Clinically significant data were reported from both click-evoked 'delayed' emissions, and distortion product emission techniques. On the basis of instrumentational complexity, click-evoked emissions were judged easier to apply to clinical use. General emphasis was placed on the need for special signal processing techniques to eliminate probe and middle ear artifacts. Use of the click-evoked acoustic response of the ear canal was recommended as a way of monitoring the coupling of the probe to the ear. Signal processing exploiting the nonlinear properties of the emission was recommended as a way to distinguish the emission from the stimulus artifact at short latencies, where the two overlapped. The paper also demonstrated how frequency analysis of the emission yields data correlated with the audiogram, and that cross-correlation techniques in both the frequency and time domains was useful in distinguishing response from noise.

Over the last 4 years several laboratories have designed and built their own 'clinical' OAE recording systems. These include the INSERM U254 Montpellier system (A. Uziel, R. Pujol), the IHR Nottingham. 'POEM' system (M. Lutman, A. Davies), the Royal Hallamshire Hospital's Medical Physics system, Sheffield (J. Stevens) and our own 'ILO88' system. These units have routinely given consistently good and useful results in neonates. Most use the original click stimulus and averaging system (fig. 1), but significantly all successful designs have paid special attention to probe design and to the presentation of data for interpretation by the operator. The consensus is that delayed OAE recording is of benefit in audiological investigation generally and particularly in neonatal screening.

The particular otoacoustic recording technique discussed in this report is based on the 'advanced cochlear echo system' proposed by Bray and Kemp [5] in 1987 as being especially suitable for infant screening. Bray and Kemp developed the recommendations of Kemp et al. [4] and, in contrast to the original simple averaging method, proposed that signal processing priority was given to increasing the certainty about whether a true emission is present or not. Bray and Kemp [5] implemented their design on a mini-computer which was not physically suitable for use in clinical service.

The design has now been refined and implemented in a form that can be installed into any IBM compatible personal computer. The mobility and reproducibility of this 'ILO88' arrangement has allowed us and other workers to gain experience with the technique in a wide variety of locations and environments. Common practical problems have been identified. The strengths and weaknesses of the technique and of OAE measurements generally have been more clearly identified.

This paper reports our initial experiences using the ILO88 IBM PC system. Our aim is to offer a very practical guide to collecting otoacoustic measurements with this kind of technique, to discuss interpretation, and to highlight future possibilities. Emerging from all this work is the need to begin setting standards for the acceptance of an otoacoustic test result as valid, and for quantifying the statistical significance of a result.

Stimulus Design and Signal Preprocessing

OAEs are continuously generated by the ear whilst exposed to sound, and sometimes even when in the quiet. That is to say a small part of the vibration of the tympanum and therefore of the sound pressure in the ear

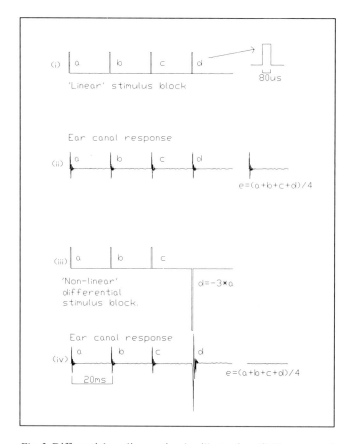

Fig. 2. Differential nonlinear stimulus illustration. (i) The ear canal probe is driven by rectangular pulses, of 80 μs duration. (ii) The train of ear canal responses can be averaged to enhance signal to noise, so that the OAE can be seen. The response waveform is unchanged by the averaging. (iii) In the nonlinear differential method used here, every fourth stimulus is inverted and three times greater in amplitude. A subaverage of the four ear canal responses contains no probe or meatal response. This assists the search for OAEs and for noise artefacts. Actually only half the saturated OAE remains due to its partial cancellation. A scale up of 2 is applied to obtain the actual OAE level. Signal-to-noise ratio is therefore reduced by the method [after ref. 5].

canal is always due to the cochlea. Being a byproduct of the hearing process gives OAEs the unique property that they can be evoked and synchronised to external sound at any level. Coming from deep within the cochlea's tuning system, they show a substantial delay in responding the external sound, around ten wave cycles. These two properties make synchronous averaging of the transiently evoked ear canal acoustic response useful as a means of retrieving and recognising the emission signal, as in figure 1.

The use of simple averaging and delay criteria for identifying the OAE is not always sufficient. For clinical purposes it is necessary to be able to distinguish between a small cochlear response and no cochlear response, in the presence of noise and an extended stimulus artifact. For increased immunity to noise and artifacts we used the subaveraged nonlinear differential stimulus method which is illustrated for click stimuli in figure 2. The method is equally applicable to tone burst stimuli. In differential stimulus sets, acoustic responses to all linearly behaving systems (i.e. nonphysiological responses) are cancelled by the summation of the response elements (four in this case). Only the saturated part of the physiological emission signal is passed to the main signal averaging unit. The technique has several advantages. Firstly, it removes all middle ear artifacts which could easily be misinterpreted as a residual OAE in a truly deaf ear (fig. 3). Secondly, it allows the trapping of patient and external noise signals at much lower levels and at short latencies. Thirdly, because of the effective artifact handling, higher levels of stimulation can be used, which increases the level of emission, and results in greater uniformity and a wider frequency range of response. The disadvantage of differential working, i.e. a small loss in backgroud signal-to-noise ratio, is rarely significant in practical, noisy clinical conditions.

Click stimuli are recommended for initial otoacoustic examinations, since they give information over a wide frequency range. Our standard stimulus was approximately 85 dB SPL peak with a bandwidth of 5 kHz. The nonlinear balancing click would in this case be 95 dB SPL peak. Due to the saturating nature of the emission response, results are not critically dependent on stimulus intensity. For adults therefore we used a fixed driving voltage level to the stimulating transducer (B-type probe). Reading the sound pressure developed in the ear canal then gives information on the meatal volume and fitting quality (see below). The intersubject variance was approximately 4 dB.

For neonates (using the E-type probe), the level of stimulator drive voltage had to be reduced by a factor of 10 to obtain the standard adult stimulation level, because of the greatly reduced meatal volume. Even so, data obtained from neonates differed markedly from adult data (see below).

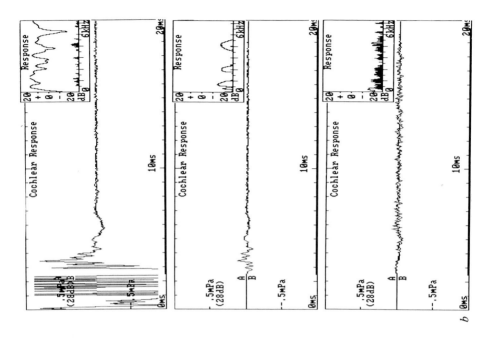

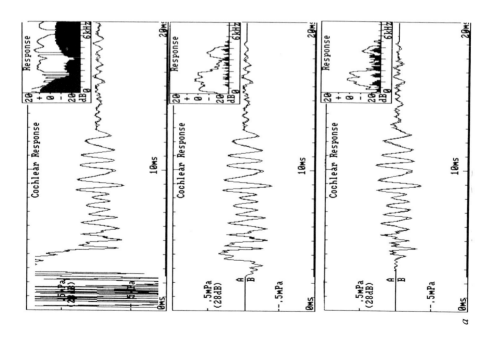

Otoacoustic Probe Design and Fitting

Fitting the probe is the single most important part of making an acoustic emission measurement. We use a compact probe design in which we try to bring the transducers within a few millimetres of the probe tip. This is to obtain good transient responses and frequency bandwidths, in a range of meatal volumes and shapes. This is important for the capture of high-frequency emissions (above 3 kHz) and for effective treatment of stimulus artifacts. Figure 4 shows the detailed design of the two sizes of probes used in our work. The larger one (fig. 4a) gives the wider response band and is preferred for all subjects except babies.

Outside of the laboratory it is often difficult to achieve an ideal fit, due to the wide variety of meatal configurations, and limited time for testing. Fitting criteria are vital and are quite different from those for tympanometry. There is less emphasis on an air-tight seal. Instead, the primary factors are the inserted frequency response and the level of noise exclusion obtained.

OAE recordings require a firm closure around a probe penetrating about halfway into the meatus. A rubber tip of appropriate size is fitted to the probe before insertion, as in typanometry. In the ear, any rubber flange on the tip should be fully compressed between the body of the probe and the canal wall. If not, noise may easily pass into the ear. The receiving and

---

Fig. 3. A comparison of the differential, and linear methods on a normal ear (*a*) and a severely impaired ear (*b*). Top row: raw averager output with a uniform click train stimulus. The 'cochlear response' panel simply shows poststimulus ear canal waveform after averaging. Two independent measurements are superimposed to confirm reproducibility. A stimulus artefact is clearly recognised from 0 to 2.5 ms because of overloading. Thereafter it is not clear where the acoustic emission begins. The response power spectrum (inset) shows the cross-power spectrum of the two responses in white, and the power spectrum of the noise, obtained by subtracting the two responses. On the top row this is meaningless due to artefact and overload. Centre row: to remove the primary stimulus artefact, a window (raised-cosine rounded over 2.5 ms) is applied to transmit signals only between 2.5 and 20 ms. Response spectra now relate meaningfully to the delayed ear canal response, which is largest for the normal ear. The deaf ear produces a small wide-band delayed response. Is this a sign of cochlear activity or not? Bottom row: performing the same operations with the nonlinear differential stimulus set resolves the issue. The deaf ear has no nonlinear (physiological) activity. Also the early part of the normal ear's response is considerably modified. This part was therefore contaminated by artefact. An increase in noise is evident with the differential method.

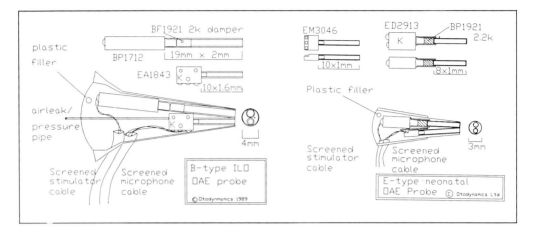

Fig. 4. Construction details for the OAE probes. Transducers and acoustic filters are from Knowles Ltd., The probes are assembled in a disposable-type plastic speculum unit. The ground lines of the stimulator and microphone transducers are not joined in the probe. The assembly is filled with setting filler plastic.

transmitting ports of the probe need to be facing the tympanum, and must not be partially screened by the rubber sealer or proximity to the meatal wall. Air-tight sealing is not a sufficient criterion, since this does not guarantee that the correct frequency response has been obtained, or that there is adequate acoustic sealing against external noise.

Figure 5 illustrates probe fitting problems found in practice, and how these are diagnosed. Figure 5a shows a good fit. The click stimulus wave-

Fig. 5. Data from one adult subject relating to six different fitting conditions with the large probe in figure 1. The stimulus driving signal is in each case a rectangular pulse of 80 µs duration, and the ambient noise level was 50 dBA. 400 stimulus presentations were summated. *a* Top row. An acceptable fit. *b* Leaky fit. *c* Poor rubber tip fitting. For *a–c* the 'cochlear response' waveform is shown and this consists of two independent measurements superimposed. The response spectrum consists of (black) the power in the difference between the two measurements which is taken to be an indication of noise contamination, and (white) the cross power spectrum of the two measurements. For the leaky fit, the noise obscures any emission and the result is useless. *d–f* Further examples. *d* Internal probe blockage. *e* Probe fallen out of the ear. *f* Good fitting, but high patient noise.

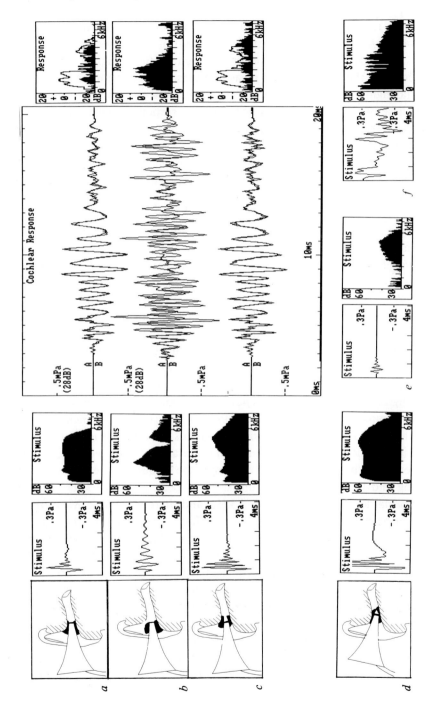

form is a short transient without much oscillation. Its power spectrum ranges from 1 to 5 kHz, which is the limit of this probe design.

A good emission waveform is seen, with negligible contradiction between the two overlayed measurements. The emission power spectrum (far right) shows noise levels (black) well below the emission level (white). These two spectra are obtained from the response-repeat difference, and the response-repeat cross-power spectrum, respectively, after Bray and Kemp [5].

In figure 5b the fit is loose. There is no seal. An oscillation of the meatal cavity in response to the stimulus produces a long ringing stimulus waveform. The stimulus spectrum shows a sharp peak (2 kHz) and trough (4 kHz), with no low-frequency energy below 1 kHz (due to the leak). A large amount of environmental noise is admitted (from the computer fan in this case) and this totally obscures the cochlear response. The response waveform is large but there is no agreement between the two superimposed responses. It is just noise and no recording is possible under these circumstances. Occasionally a large meatus can produce this kind of ringing ear canal response by virtue of its own characteristics without there being a leak. However in this case there will be stimulus energy below 1 kHz. In general, however, whenever a sharply peaked, or notched stimulus spectrum is seen, a refitting should always be tried.

The 'extended tip' problem illustrated in figure 5c can also produce ringing, but of a higher frequency. With probes having the microphone and transmitter ports in close proximity, it is important not to enclose the tip in any way. If the rubber probe tip extends beyond the end of the probe high-frequency acoustic ringing can occur around the tip cavity, and interfere with the recording. In the example the stimulus waveform shows ringing and the stimulus spectrum shows a peak around 4 kHz. There is a normal level of energy below 1 kHz, and there is no leak. Measurements can be made. The resonance peak enhances reception of the higher frequency emissions, but can also enhance artifacts. The Knowles magnetic receiver (1712) used for our probe was found to be somewhat nonlinear around 4 kHz, so that long acoustic oscillation (ringing) due to tip extension might be confused with a true emission. In general high-frequency differential OAEs should be examined very critically if there is a stimulus resonance.

A more extreme type of obstruction occurs when the probe tip and sealing rubber is pressed against the meatal wall. This is usually because of a wrong angle of entry. Figure 5d illustrates this. Typically the stimulus waveform is unusually large and has a slow (low frequency) tail due to a response of the transmitter transducer in the new small volume acoustic environment. The stimulus spectrum is unusually strong, but no OAE recording is possible. Two further negative conditions are shown in figure 5. In figure 5e, the probe has falled out of the ear. This is characterised by a very small-sized stimulus waveform. In figure 5f the probe is well fitted but the subject coughs causing a noisy stimulus display.

The above demonstrates that meatus response monitoring can be extremely useful in detecting poor probe fitting conditions, prior to OAE measurement. Fitting waveforms are modified in the case of neonates.

Adult versus Neonate Ears

We have attempted to compare adult and neonate ears using this method. We find consistent differences, illustrated by the typical examples shown in figure 6. The main differences are not attributable to the different probe designs used. The very different meatal and tympanic configurations of neonate and adult ears are thought to have a major influence. All neonate emissions are stronger than the average adult response by more than 10 dB. This may well be due to the enhancement of coupling between tympanum and microphone caused by the very small meatal volume but increased intrinsic emission activity cannot be ruled out at this stage. Neonatal emissions typically extend fairly uniformly from 1 to 5 kHz, whereas adult responses very often have less power at high frequencies and more below 1 kHz. Adult responses almost always have 'missing' frequency bands or notches, which are not usually associated with significant hearing loss. Neonatal responses rarely have deep notches. We interpret the difference as an aging effect.

Examination of the stimulus waveform and spectrum recorded in the neonate shows the characteristic of resonance as in the extended tip fitting example in figure 5c. It also shows a loss of low frequencies as in the leaky fit example in figure 5b. In practice, most neonatal fittings tend to be slightly leaky, as a precaution. Also even when sealed, the meatal tissue is so thin that it presents little impedance to the transmission of low frequency sound. This also means that environmental noise must be kept lower for neonates than for adults, although this is not often a problem. The resonance at 4 kHz with the E-type probe was found in most neonates and seems to be a property of the probe-middle ear coupling. The probe

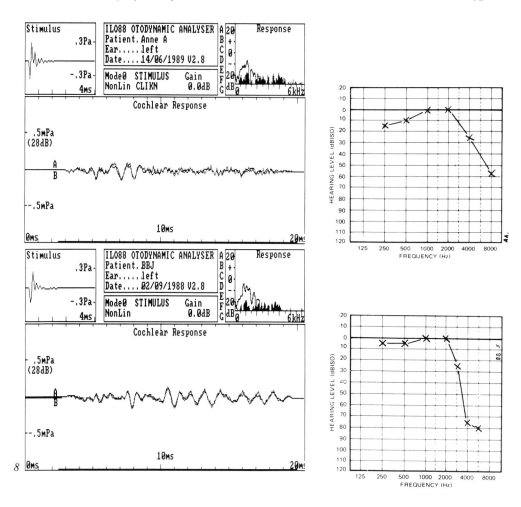

Fig. 8. Emission from two patients with high-frequency loss. Note the low frequency emissions.

Fig. 9. Data from one ear obtained with a tone burst of 4 waves cosine windowed. Top left is for 500 Hz (stimulus longer than stimulus panel), bottom left for 1,000 Hz, and top right, for 2,000 Hz. Bottom right: the 500 and 1,000 Hz stimuli are played simultaneously. This evokes a response just like the summation of the two individual responses. This is an additive, linear behaviour.

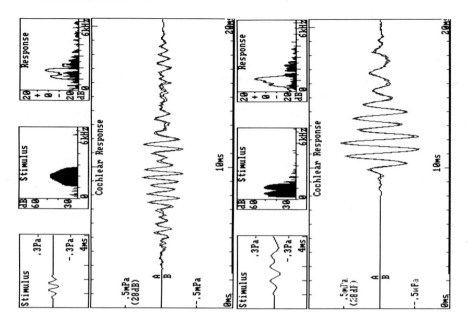

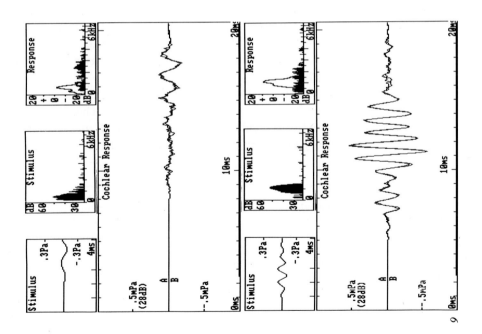

Practical Uses of Tone Burst Stimuli

Tone burst stimulation can be effectively used with the nonlinear differential technique and so can yield reliable emission data. Tone burst stimuli have a number of advantages. It is not however time effective to make a series of separate measurements with a set of tone burst frequencies just to screen for normal activity. A wideband click evoked measurement should be completed first. The special advantage of tone pulse stimulation does not lie in its frequency specificity as such, since equal frequency specificity is available by analysis of click-stimulated responses (see above). Rather, tone bursts allow a greater concentration of energy to be applied to regions of the cochlea without reaching the nonlinear overload region of the transducers. In other words, the increased peak intensity is exchanged for duration. Outside of the most active otoacoustic band (1–2 kHz), responses can sometimes be elicited by tone burst stimulation at where they could not be clearly identified with click stimuli.

Quantification and Standardisation

There is a growing need for common standards and practices for OAE measurements. Technically, the test is more like typanometry than either subjective or evoked response audiometry. Calibration of the probe and amplification are not themselves difficult tasks but it will be necessary to agree on procedure, and on grades of bandwidth and linearity and other aspects of acoustic performance.

Regarding interpretation of the measured data, a final audiological conclusion on the meaning of a test result has to be approached in four stages. These are: Is the fitting good? Is the recording valid? Is an emission present? Is the auditory periphery behaving normally? We have already

---

Fig. 10. Data from another ear (as for fig. 6, bottom left) exposed to tone bursts of 1 kHz (top left), 3.6 kHz (bottom left) and 1 and 3.6 kHz simultaneously (top right). Note that the 3.6 kHz stimulus excites a sustained emission (possibly spontaneously active). Bottom right: the algebraic sum of the individual 1 and 3.6 kHz responses. This exactly matches top right data, and demonstrates linear behaviour between these well-separated frequency bands. The centre inset spectrum shows the click-evoked power spectrum dotted. Note that the major difference is where the energy is missing from the tone burst combination.

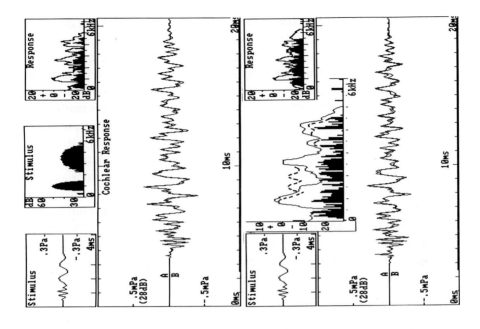

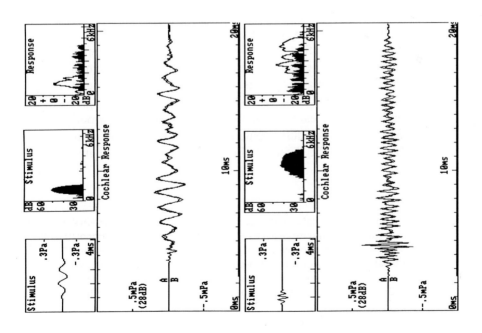

discussed the value of stimulus waveform monitoring. To assess the validity of a recording it is useful to quantify the noise contamination of the delayed response, and this is obtained from the difference between two repeat (interleaved measurements). A test should be rejected if emissions are not seen above the noise and if the noise itself is strong enough to obscure normal small responses. The energy in the response is easily computed from the waveform. The reproducibility is conveniently expressed as the correlation between the two waveforms. Low waveform correlation may mean noise contamination. This may not mean the emission is missing, or that the test result should be rejected. The use of frequency analysis and cross correlation allows the signal and noise to be assessed at each frequency. Often an overall poor waveform correlation is a result of low-frequency noise, and the high-frequency emission can be clearly seen on the waveform and response spectrum. Only by an assessment of the emission and noise power as a function of frequency can an audiological conclusion be drawn. The data in figure 7 includes the numerical analysis normally generated and assessed. The noise panel records the number of samples taken and rejected, and the level of the mean contamination in the data accepted. The response panel shows the echo dB SPL mean level, the correlation of the repeat measurement and the dB SPL of the difference, which is the noise signal in the final result. Finally, the stimulus peak level and beginning-end correlation is quoted, together with the time of testing. All these help in the assessment of the test result.

Figure 11 shows a first step towards normative data for emission energy. It allows a percentile figure to be obtained for a patient's emission level at any frequency, using our standard click stimulus. The general trend is for levels above 3 kHz to be 10 dB lower than below this frequency. This may well be a middle ear transmission property. Note that around 3 kHz, more than 10% of the adult population did not produce an emission sufficiently strongly to be detected above the noise. The neonate response

---

Fig. 11. Normative data. Top figure shows the nonlinear differential emission energy level (dBSPL/50 Hz) exceeded by 10, 30, 50, 70, and 90% of a population of 24 normally hearing subjects between 18 and 30 years, as a function of frequency. Centre: a typical neonate response superimposed on the 10 and 90 percentile data. Bottom: data from a high-frequency hearing loss patient. Note: with calibration standards yet to be agreed for otoacoustics, these data should be used only as a guide for the collection of local normative data.

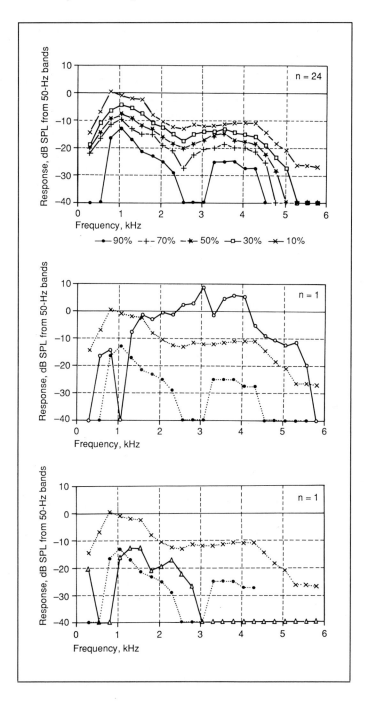

clearly stands above the best adult level from 1.5 kHz upwards. The high frequency loss patient is seen to be in the bottom 30% of the population up to 3 kHz, and beyond the bottom 10% in his hearing loss region.

Because of the lack of agreed calibration standards these data should *not* be used for clinical interpretation even with ILO88 instruments, but rather as a model with which to design local normative data pools. On a practical level it may be worth repeating that when averaging substantially different decibel values, errors will occur unless computations are performed on the true pressure or power (pressure squared) value as appropriate. Similarly measured noise levels should not be subtracted from emission levels in the decibel form, or serious error will result.

Conclusion

We have attempted to discuss important practical aspects of acoustic emission measurement and to illustrate these wherever possible. Earlier uncertainties about the interpretation of emission have largely been resolved, and several types of instrumentation are now entering wider use, including our own ILO88 unit. It is increasingly apparent that standards are needed for the calibration and interpretation of OAE data and for the exchange of information between centres. Regarding the future, it would seem that otoacoustic testing is ready to be introduced into neonatal screening. It also has a role in investigative audiology.

References

1   Kemp DT: Hearing faculty testing apparatus. European patent EP0015258. 1979.
2   Kemp DT: Stimulated acoustic emission from the human auditory system. J. Acoust Soc Am 1978;64:1386–1391.
3   Johnsen NJ, Elberling C: Evoked acoustic emission from the human ear. III. Neonates. Scand. Audiol 1982;12:17–24.
4   Kemp DT, Bray PB, Alexander L, Brown AM: Acoustic Emission Cochleography – Practical Aspects. Scand Audiol 1986;Suppl. 25:71–94.
5   Bray PJ, Kemp DT: An advanced cochlear echo suitable for infant screening. Br J Audiol 1987;21:191–204.

David T. Kemp, PhD, Functional Analysis Laboratory, Audiology Department, Institute of Laryngology and Otology, UCMSM, 330 Gray's Inn Road, GB–London WC1X 8EE (UK)

Grandori F, Cianfrone G, Kemp DT (eds): Cochlear Mechanisms and
Otoacoustic Emissions. Adv Audiol. Basel, Karger, 1990, vol 7, pp 99–109

Deconvolutions of Evoked Otoacoustic Emissions and Response Nonlinearity

F. Grandori, P. Ravazzani

CNR, Centro Teoria dei Sistemi and Dipartimento di Elettronica
Politecnico di Milano, Italia

Clinical applications of evoked otoacoustic emissions (EOAEs) are limited by several factors. First of all, EOAEs vary greatly from ear to ear, even within the same subject, for delay, rise and decay time, overall duration and temporal fine structure or frequency content [1–6], despite the fact that, for a given ear and with the same input stimuli, the reproducibility is surprisingly high, even in responses recorded several years apart [7].

A second serious limit is constituted by the presence in the recorded signal of an initial portion contaminated by a variety of acoustical reflections lasting several milliseconds. The duration of these artifacts depends strongly on the type of stimulus used to evoke the responses, the overall acoustical characteristics of the probe, the volume of the external meatus and the coupling between probe and external auditory canal.

The presence of these artifacts poses a serious limit to unambiguous detection of the emission into the recorded data samples.

The aim of this paper is to investigate some interesting properties of the generating mechanisms of otoacoustic emissions: linearity of the responses, as revealed by different stimuli (clicks and tone bursts), at a given intensity level, and the deviations from the linearity that are observed in the responses evoked at different intensity levels. It will be shown how these properties can be used to discriminate, at least theoretically, between otoacoustic emissions and artifacts.

These aspects are studied with a method of deconvolution (inverse filtering) of the otoacoustic emissions, in normal ears. A similar approach was previously suggested by Elberling et al. [8] and the present author [9]. Some preliminary results were presented elsewhere [10].

Material and Methods

EOAEs were measured with two different systems. In a first series of experiments, an Amplaid miniature probe containing an electroacoustic transducer and a microphone was used. Details of the probe and of the recording procedures are described elsewhere [5–7, 11, 12] and summarized herein. Clicks (0.1 ms duration) and tone bursts of various frequencies (500, 1,000 and 2,000 Hz, 1-ms rise and decay time, with a plateau of 2 ms) at a rate of 21/s were delivered at different intensity levels (0–50 dB HL). To achieve a good signal-to-noise ratio, responses within the frequency band 200–5,000 Hz were averaged over 1,500–2,000 repetitions within a time window of either 20 or 30 ms (512 points), with a prestimulus interval of 2 ms. Stimulus generation and data recording were under the control of an Amplaid MK 7 system.

In a second series of experiments, an ILO88 system was used in the linear mode of operation. Click stimuli were used with intensities ranging approximately from 60 to 90 dB p.e. SPL (sampling frequency 25 kHz, 512 points). With this system, a larger recording bandwidth was allowed (upper limit around some 5–6 kHz).

In both cases, data manipulations were done with a host computer at the Polytechnic of Milan.

Processing Techniques

In the first experiment, responses were deconvoluted with the input waveforms (clicks and tone bursts), as recorded from the probe itself in situ, at three different stimulus intensities. These signals are then compared with the click responses to extract some information about the linearity of the system at a given intensity level.

For this set of data (Amplaid system), data processing included:

(1) Off-line zero-phase band-pass digital filtering (Butterworth, 400–2,500 Hz, 60 dB/octave);

(2) Deconvolution of the responses within the time window 8–28 ms post stimulus time (pst);

(3) Further digital zero-phase low-pass filtering to remove unavoidable high-frequency noise due to the deconvolution technique [13].

For the second series of recordings (ILO88), data were transcoded and analyzed on a host computer. With this system, only responses to clicks (linear mode) were considered at several different intensity levels and compared to each other to investigate the changes of the deconvolutions with time pst and intensity.

Basically the same parameters were used as in experiment 1, a part from the bandwidth (low-pass cut-off frequency of 5 kHz). The deconvolution process was initiated at 3.2 ms, that is to say that the initial portion 0–3.16 ms was not considered. Data were windowed (Kaiser window) 3.2–4.76 and 18.92–20.48 ms; after deconvolution, a Butterworth filter was used (400–4,000 Hz).

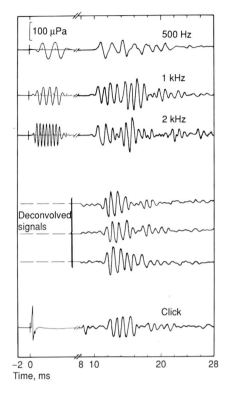

Fig. 1. An illustrative example of the deconvolution procedure. The first three traces from top illustrate the responses to tone bursts of different frequencies; all responses were recorded at the same dB HL (inserts are the initial transients and are plotted with an arbitrary scale). The second set of three traces shows the results of the deconvolution procedure for the traces above. The bottom trace is the response to click for the same ear.

Results

Experiment 1

EOAEs from both ears of 4 subjects at intensities of 10, 20 and 30 dB HL were analyzed as described above. The results can be summarized as follows: for a given ear, at a given intensity level, deconvoluted signals at the different burst frequencies are quite close to each other (maximum correlation coefficient in the range of 80–90%) and very close to the click response evoked at comparable intensity levels. Some examples of the responses and of their deconvolutions are illustrated in figure 1, from a representative subject, together with the click response. Other examples

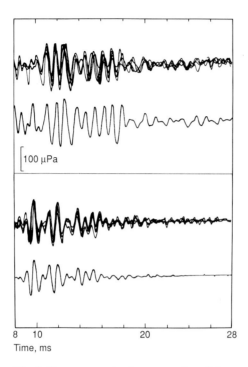

8   10   20   28
Time, ms

Fig. 2. The two panels show examples of deconvolutions from two subjects with 'long' and 'short' responses, respectively, with the same procedure as illustrated in figure 1. In each panel, the deconvolutions of the responses to tone bursts are superimposed; click responses are given below. It is important to note how close the deconvolutions are to each other and how close they are to the click response from the same ear.

from two other ears are given in figure 2. To stress the similarities existing among the deconvoluted waveforms, the signals were superimposed.

Some subtle differences were observed among the deconvolutions at the different intensities, and among these and the click responses at corresponding intensities, in particular for the rate of decay of the deconvoluted signals at the lowest intensities. Due to the limited number of conditions examined, however, no systematic attempt was made to quantify these aspects.

Experiment 2

Responses were examined at several different intensity levels. Examples of stimuli, emissions and deconvolutions are shown in figure 3. It is

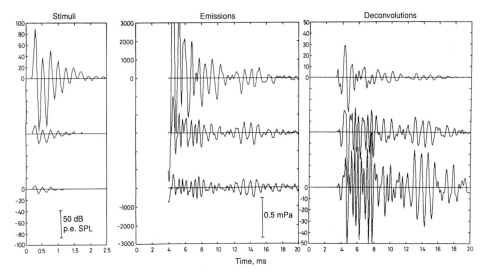

Fig. 3. The principle of the deconvolution between the input stimulus (left) and the emissions (center). Deconvoluted signals (right) are given in arbitrary units. Note the increase in the amplitudes of the deconvolutions at the lowest intensities.

seen that at the various intensity levels the emissions are only slightly different in amplitude, due to saturation mechanisms [1, 5]. By contrast, the deconvolutions vary greatly from the highest to the lowest intensity level. In figures 4 and 5 other examples of complete series of emissions and deconvolutions are illustrated, from four normal ears.

As observed in figure 3, the overall dynamic range is higher for the deconvolutions than for the emissions, and the deconvolutions of responses at the lowest intensities are larger. This is simply due to the process of deconvolution (a linear operation) and confirms the nonlinear character of emissions.

To explore in more details this aspect, the rms amplitude of the responses were measured in 6 different time windows (fig. 6). Duration of each window is 75 sample data points (3 ms); analysis starts at 3 ms.

What is clear from figure 6 is that the relationship between stimulus intensity and response amplitude strongly depends on the time pst. Note that for the first two to three time windows, approximately from 6 to a maximum of 12 ms, the slope is much higher than for the latest windows, for all the subjects.

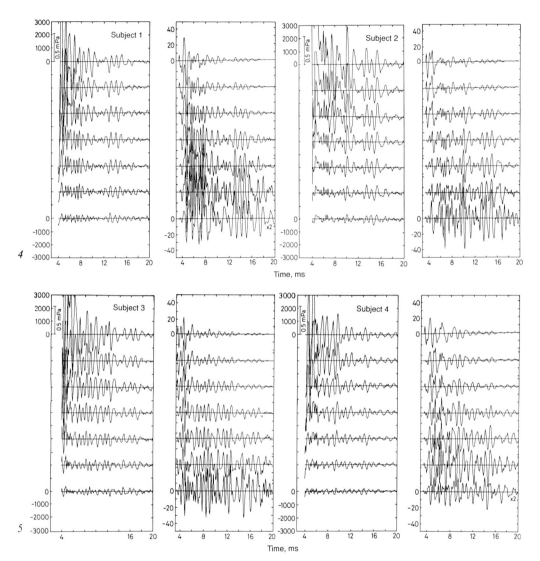

Fig. 4. Example of emissions (left) and corresponding deconvolutions (right) from two representative subjects. Stimulus intensities decrease from top to bottom (from about 96 to 63 dB p.e. SPL). EOAEs are plotted from 4 ms. For clarity, the initial portions of the recorded signals are not represented. Deconvolutions are computed and plotted from 3.2 ms.

Fig. 5. Another example of EOAEs and deconvolutions from two other subjects. Details as in figure 4. Note again the high degree of reproducibility of the waveforms of deconvolutions as a function of stimulus intensity.

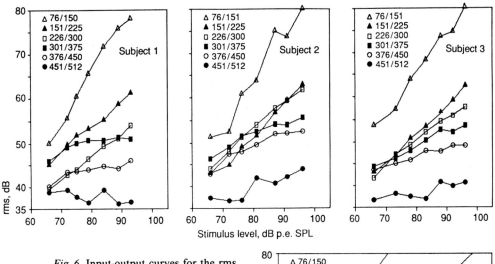

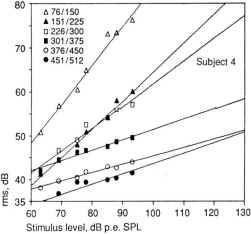

Fig. 6. Input-output curves for the rms magnitude of EOAEs from four subjects, in decibels relative to an arbitrary value (unity). Data were computed for six different time windows; each window is identified by its own initial and final sampling points. In the bottom panel, data have been linearly interpolated to stress the differences in the slopes of each input-output curve. Note that the slopes of the curves for the first two to three windows are much steeper than those at the greatest latencies. This suggests that saturating mechanisms are much stronger at the greater latencies, while the earlier portion of the responses exhibits a linear behavior with a gain close to unity (see also fig. 7).

For each experimental curve, data have been interpolated with a linear regression line. When slopes of the regressions are plotted (fig. 7), this trend is clear. A horizontal line with a slope of 1 is plotted in the same figure to visualize the input-output characteristic of a system with a unity gain. Results of figures 6 and 7 show that, at the shortest delays, responses are characterized by a slope close to unity, while at the longest pst intervals the responses are increasingly saturated. This trend has already been

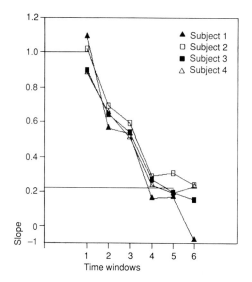

Fig. 7. Slopes of the input-output curves shown in figure 6. Note that the slopes drop from a value close to unity (at the shortest pst) to a value about four to five times lower.

observed in some preliminary investigations into the input-output relationships of EOAEs [5, 11].

A similar analysis of the deconvolutions of the same responses reveals basically the same phenomena (fig. 8, 9). As noted earlier, rms values of deconvolutions tend to increase as the intensity is decreased (fig. 8).

Conclusions

Results presented here confirm some characteristics of EOAEs which have been observed in other related experiments, like linearity at a given intensity level of the stimulus-response relationship. This seems to confirm the view that EOAEs are generated by discrete sources distributed along the cochlea; when excited by stimuli of comparable intensity levels, the responses can be described as produced by a linear system. This simplistic view, however, is no more valid when responses at various intensities are considered. In this case, the impulse responses that can be estimated at a

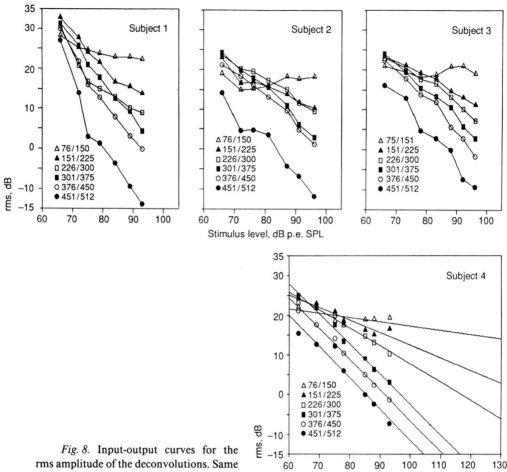

Fig. 8. Input-output curves for the rms amplitude of the deconvolutions. Same subjects as in figure 6.

given intensity level is similar in waveshape throughout the intensity range, but its magnitude changes with the stimulus level, reflecting the intrinsically nonlinear character of the generating mechanisms.

At the lowest intensities investigated in the present experiments, in any case well above hearing thresholds, the waveform of the deconvolutions remains substantially constant, while the magnitude is increased. This suggests the possibility of developing a new method for detecting the

Grandori F, Cianfrone G, Kemp DT (eds): Cochlear Mechanisms and
Otoacoustic Emissions. Adv Audiol. Basel, Karger, 1990, vol 7, pp 110–116

Spectral Line Width of Spontaneous Otoacoustic Emissions

H.P. Wit, P. van Dijk

Institute of Audiology, University Hospital, Groningen, The Netherlands

Already at the first symposium fully devoted to active mechanical processes in the cochlea (London, September 1979), the possibility of 'self-sustaining action' and 'build up of continuous oscillation' was put forward by Kemp [1] and Wilson [2], in order to describe what was, at that time, called 'objective tinnitus'.

Johannesma [3] was the first who proposed the Van der Pol equation for the generator of spontaneous otoacoustic emissions (SOAEs). To this type of oscillator, energy is supplied for small values of the descriptive variable x(t) (displacement in a mechanical system), while power is dissipated for larger x. The result is a limit cycle oscillation. The limit cycle being a closed trajectory of the point $(x, \dot{x})$ in phase space [see, for instance, the tutorial paper by Hanggi and Riseborough, 4].

The mathematical form of the differential equation for an undriven Van der Pol oscillator is:

$$\ddot{x} - \gamma (1 - x^2) \dot{x} + \omega_o^2 x = 0.$$

For small values of the parameter γ (compared to ω_o) the limit cycle is almost a circle [4], indicating that the solution is an almost perfect harmonic oscillation (with amplitude 2).

For larger values of γ higher-order harmonics must be added to the solution [5]. Such higher harmonics have not been observed in recordings of SOAEs, so we expect an SOAE signal to be a pure sinusoid, to which microphone noise is added.

Careful analysis shows that this is not the case [6]: amplitude and phase of an SOAE signal show fluctuations. Part of these fluctuations are indeed due to noise of the measuring microphone; but another part is intrinsic oscillator noise: thermal noise being associated with the damping and noise of the amplifier in the feedback loop [6].

Especially the phase-locking behavior of an SOAE to an externally applied periodic force can satisfactorily be described only by adding noise to the oscillator. This noise brings about measurable departures from completely locked-in behavior [7] if the externally applied force is not too strong [a theoretical description of a driven nonlinear oscillator in the presence of noise is given in ref. 4].

If a small deviation is given to the amplitude of a Van der Pol oscillator (for instance by a shortly acting external force), the amplitude relaxes back to its limit cycle value. The relaxation time for this process is determined by the parameter γ in the Van der Pol equation [4, 6, 8]: $\tau_{rel} = \gamma^{-1}$. Small amplitude deviations of an SOAE signal are continuously brought about by the permanently present intrinsic noise force.

By measuring τ_{rel} for an SOAE signal it would be possible to obtain the parameter γ. Direct measurement of τ_{rel}, however, is difficult. Therefore we did not measure the amplitude relaxation time τ_{rel}, but the correlation time τ_A for amplitude fluctuations.

It can be shown that τ_A equals τ_{rel} if the spectral bandwidth of the oscillator noise force is much larger than the oscillator frequency. We assumed this to be true. The time τ_A was measured by filtering an SOAE signal in a narrow band (6 or 12% bandwidth) around the emission frequency. The envelope of the resulting signal was obtained after half-wave rectification and low-pass filtering; and with an FFT spectrum analyzer, the frequency spectrum of this envelope was produced [6]. A typical result for a human subject is given in figure 1. The solid line in this figure is a computer fit to the FFT analyzer output, being the sum of a Lorentzian and a constant term. The constant term accounts for background noise in the SOAE signal, mainly produced by the measuring microphone, while the Lorentzian peak is the result of (low frequency) amplitude fluctuations of the SOAE signal itself. By measuring the width of this peak, τ_A is obtained. Table 1 gives values for τ_A for five different SOAEs (from five different human subjects).

By measuring the area under the Lorentzian in figure 1, the relative contribution of the SOAE amplitude fluctuations to the RMS amplitude of the above-mentioned envelope could be calculated. This RMS amplitude was directly measured and the RMS value δA_{rms} of the SOAE amplitude fluctuations was derived from the results of this measurement. Values for

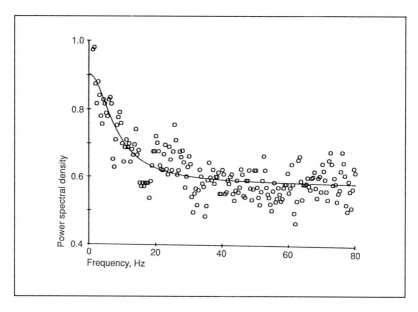

Fig. 1. Frequency spectrum of envelope of bandpass-filtered SOAE signal. The Lorentzian peak around $f = 0$ accounts for amplitude fluctuations caused by intrinsic SOAE oscillator noise.

$\delta A_{rms}/A_o$ obtained in this way are also given in table 1 (A_o is the average amplitude of the SOAE signal).

If a Van der Pol oscillator is subject to an intrinsic noise force, not only the amplitude of the oscillator signal fluctuates, but also the phase. The rate D_φ of these phase fluctuations is determined by the power spectral density of the noise force. Also δA_{rms} depends on this spectral density, and on the parameter γ in the Van der Pol oscillator equation. Because also τ_{rel} depends on γ it can be shown that D_φ, $\delta A_{rms}/A_o$ and τ_{rel} (or τ_A if the noise is broadband) are related [6] according to:

$$D_\varphi = (\delta A/A_o)^2/\tau_{rel}.$$

This relation predicts a value for D_φ, based upon the measured values for $\delta A_{rms}/A_o$ and τ_{rel} ($= \tau_A$). Predicted D_φ values are given in the third column of table 1.

It is also possible to measure D_φ, because the phase fluctuations caused by the oscillator noise manifest themselves through broadening of

Table 1. Correlation time for amplitude fluctuations, relative amplitude fluctuations, predicted and measured phase diffusion constant for SOAEs from 5 different human subjects

Subject	τ_A, ms	$\delta A_{rms}/A_o$	D_φ, s^{-1}	
			predicted	from line width
WK	20	0.021	0.02	2.8
MA	15	0.028	0.05	2.0
RL	15	0.063	0.26	9.4
MD	11	0.020	0.04	3.3
CS	10	0.017	0.03	2.5

the SOAE spectral peak. In the standard mode of the FFT analyzer the linewidth of a human SOAE spectral peak falls within one channel. Therefore the SOAE signal was multiplied by a fixed sine wave with a frequency slightly below the SOAE frequency and low-pass filtered before spectral analysis. This 'zoom' procedure improves the resolution in the frequency domain (if a long enough sample of the SOAE signal is available). A Lorentzian was fitted by computer to the output signal of the FFT analyzer. Results are given in figure 2. This figure shows that the spectral line width of a human SOAE is of the order of 1 Hz.

The relation between the spectral line width Δf (full width at half maximum in the power spectrum) and the phase diffusion constant D_φ is given by [8]: $D_\varphi = 2 \pi \Delta f$. values for D_φ derived from the spectral line width are given in the last column of table 1.

These measured D_φ values are two orders of magnitude larger than predicted D_φs (third column table 1), which means that the Van der Pol oscillator with broadband intrinsic noise does not give an adequate description for an SOAE. This statement is contradictory to the conclusion drawn in one of our earlier papers [6]. But in that work D_φ was derived from fluctuation in the intervals between successive zero crossings of the emission signal. Apparently, this procedure yields too small a value for D_φ, because these intervals are not long compared to the correlation time of the noise. This means that the phase obtained with this method was wrongly assumed to have a 'diffusional character' (if the phase has diffusional character the variance of the phase shift is directly proportional to time; the proportionality constant being D_φ [8]).

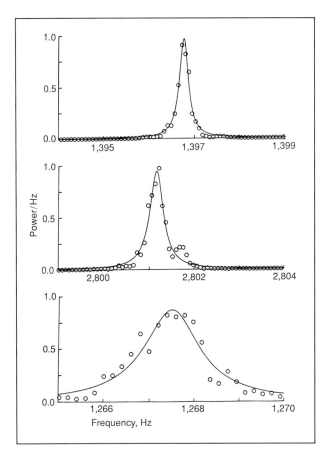

Fig. 2. Spectral peaks of SOAE signals from three different human subjects. Solid lines are Lorentzians fitted by computer. Peak widths are 0.25, 0.37 and 1.51 Hz from top to bottom.

There are several explanations for the discrepancy between the predicted and measured value of D_φ:

(1) If the oscillator intrinsic noise is not broadband, the amplitude relaxation time τ_{rel} cannot be replaced by the amplitude correlation time τ_A. It can be shown that for narrow band noise $\tau_{rel} = \tau_A - \beta^{-1}$, in which β is the noise bandwidth. So if β^{-1} is approximately equal to τ_A, then τ_{rel} will be much smaller than τ_A, leading to a much larger predicted value for D_φ. But to make the predicted value for D_φ equal to the measured value, τ_{rel} has to

be a few tenths of a millisecond (if we assume that the measured values for $\delta A_{rms}/A_o$ are correct). This in turn would mean that γ (being equal to τ_{rel}^{-1}) divided by ω_o would have a value of approximately 1, and the oscillator limit cycle would deviate markedly from a circle [9]. (The parameter ε, used by Stoker [9], is equal to our γ/ω_o. As many theoreticians, Stoker has set ω_o equal to 1 by a time scale transformation.) The consequence of this deviation would be a third harmonic in the SOAE frequency spectrum less than 20 dB below the fundamental. Such higher harmonics have never been reported thus far for SOAEs.

Another argument against narrow-band intrinsic oscillator noise can be derived from the time constant for change of SOAE amplitude after suppression with an externally supplied tone. Schloth and Zwicker [10] give an average value of 13 ms for this time constant. Although suppression phenomena in a Van der Pol oscillator have not been thoroughly investigated theoretically, it is obvious that this time constant and τ_{rel} are closely related. This gives a value for τ_{rel} with the same order of magnitude as the measured values for τ_A and leads to the conclusion that the noise is broadband.

(2) Another possibility is that the parameters that determine the frequency of an SOAE show small random fluctuations. If this frequency is, for instance, determined by a mechanical stiffness, fluctuations of the stiffness parameter κ would lead to a broadening of the SOAE spectral peak and thus to a larger than predicted D_φ value.

(3) Intriguing is the possibility that the stiffness parameter κ does not fluctuate randomly, but is a nonlinear function of the place variable x.

If, for instance, $\kappa = \kappa_o + \beta x^2$, and we assume for simplicity that no resistance is present, then the differential equation for the motion of a particle with mass m becomes: $m\ddot{x} + (\kappa_o + \beta x^2) x = 0$. This is the equation for the well-known Duffing oscillator [9]. Its special property is that its frequency depends on its amplitude. (For $\beta > 0$, 'hard-spring condition', frequency increases with amplitude, while for the 'soft-spring situation', $\beta < 0$, it decreases with increasing amplitude.) By combining Duffing's and Van der Pol's equations, we obtain a self-sustained oscillator with a nonlinear stiffness term:

$$m\ddot{x} - \gamma_1 (1 - x^2) \dot{x} + (\kappa_o + \beta x^2) x = 0.$$

that somewhat more elegantly reads:

$$\ddot{x} - \gamma (1 - x^2) \dot{x} + \omega_o^2 (1 + \alpha x^2) x = 0,$$

with $\omega_0^2 = \kappa_0/m$. If such an oscillator is subject to a noise force, the noise force will induce (low frequency) amplitude fluctuations, leading to frequency fluctuations and consequently to broadening of the SOAE spectral peak. But if this is true, SOAE amplitude and momentary frequency (time between zero crossings) are highly correlated. Experiments to investigate whether such a correlation is present, are in progress in our laboratory.

In conclusion, we can say that SOAEs can in first approximation satisfactorily be described by a Van der Pol oscillator. The too large spectral line width of an SOAE, however, is in conflict with this model. There are several possibilities for a more precise description of the SOAE oscillator, but at this moment there are not enough experimental data to decide between these possibilities.

References

1   Kemp DT, Chum R: Properties of the generator of stimulated acoustic emissions. Hear Res 1980;2:213–232.
2   Wilson JP: Model for cochlear echoes and tinnitus based on an observed electrical correlate. Hear Res 1980;2:527–532.
3   Johannesma PIM: Narrow band filters and active resonators; in van den Brink G, Bilsen FA (eds): Psychophysical, Physiological and Behavioural Studies in Hearing. Delft, Delft University Press, 1980, pp 62–63.
4   Hanggi P, Riseborough P: Dynamics of nonlinear oscillators. Am J Physics 1983;51:347–352.
5   McLachlan NW: Ordinary Differential Equations in Engineering and Physical Sciences. Oxford, Oxford University Press, 1950.
6   Bialek W, Wit HP: Quantum limits to oscillator stability: Theory and experiments on acoustic emissions from the human ear. Physics Lett 1984;104A:173–178.
7   Van Dijk P, Wit HP: Phase-lock of spontaneous oto-acoustic emissions to a cubic difference tone; in Duifhuis H, Horst JW, Wit HP (eds): Basic Issues in Hearing. London, Academic Press, 1988, pp 101–105.
8   Stratonovich RL: Topics in the Theory of Random Noise. New York, Gordon and Breach, 1981, vol 2.
9   Stoker JJ: Nonlinear Vibrations in Mechanical and Electrical Systems. New York, Wiley, 1950.
10 Schloth E, Zwicker E: Mechanical and acoustical influences on spontaneous oto-acoustic emissions. Hear Res 1983;11:285–293.

H.P. Wit, PhD, Institute of Audiology, University Hospital, P.O. Box 30.001, NL–9700 RB Groningen (The Netherlands)

Grandori F, Cianfrone G, Kemp DT (eds): Cochlear Mechanisms and
Otoacoustic Emissions. Adv Audiol. Basel, Karger, 1990, vol 7, pp 117–125

Methods and Preliminary Results of Measurements of Distortion Product Otoacoustic Emissions in Normal and Pathological Ears

R. Probst, C. Antonelli, C. Pieren

Department of Otorhinolaryngology, Kantonsspital, University of Basel,
Switzerland

Distortion products (DPs) are general phenomena of many physical systems. They are generated by nonlinear elements which distort the signal and thereby create additional frequencies when excited by sinusoids. The presence of DPs in the human auditory system has been known for many years [1, 2]. Later, mainly psychoacoustical measurements [3–5] demonstrated the existence of DPs, and thus of nonlinear elements, in the auditory system at medium level stimuli. Goldstein [5] proposed the basilar membrane as the locus of generation.

Shortly after the discovery of otoacoustic emissions (OAEs) [6], such emissions linked to DPs (DPOAEs) were demonstrated in human ears [7]. They were generated by pure tones and consisted of acoustic energy above the noise floor detectable in the spectrum of the ear canal at frequencies other than the stimuli. The stimulus frequencies are called the primaries ($f_1 < f_2 < f_3$, etc.), and the frequencies of DPs are related to them by exact mathematical formulas. In human ears, as well as in the ears of many animals species, the difference intermodulation DP at the frequency of $2f_1 - f_2$ is the most prominent OAE, and is therefore the most thoroughly examined. The amplitude of a DPOAE is dependent on the amplitudes of the primary stimuli. While many rodents exhibit DPOAEs of approximately 40 dB smaller than the stimulus levels, in human ears they differ by 60 or more dB.

This report summarizes our first experience with relatively simple methods of recording DPOAEs in normal and pathological human ears.

Materials and Methods

Subjects

Subjects without any otological complaints and patients with hearing losses due to various pathologies were examined. Seventy-seven ears of 46 subjects were classified as normal. The definition of normal included a negative otologic history and examination, no medication, normal findings at impedance audiometry, thresholds of 20 dB HL or less at all standard audiometric frequencies (0.25, 0.5, 1, 1.5, 2, 3, 4, 6, 8 kHz), and an average of these thresholds of 10 dB HL or less. Thirty-six ears of 25 subjects were classified as 'near-normals'. The subjects were without symptoms and met the same criteria as the normals with the exception of their audiometric thresholds. These were 40 dB HL or less a the standard audiometric frequencies and the average was 20 dB HL or less. Fourteen ears of subjects with 'normal' hearing on the other side were included in this group.

The patient group included 44 subjects with cochlear hearing losses of various pathologies, such as Menière's disease, noise-induced hearing loss, or sudden hearing loss. Middle-ear disturbances or retrocochlear involvements were audiometrically excluded with impedance audiometry and/or evoked brainstem potentials. Eighty-six ears were tested. Some of the patients had near-normal hearing on one side. However, these ears were included in the patient group as well. They were not included in the near-normal group, since the patients complained about hearing disorders at least on one side.

Testing was carried out in a sound-proof booth. The measurement of DPOAEs took about 10–15 min per ear.

Stimuli

Two continuous tones at two frequencies, f_1 and f_2, were generated by a two-channel frequency synthesizer (Hewlett Packard 3326A). They were delivered to a pair of insert earphones (Etymotic Research ER-2) which were connected by calibrated tubes to an

Table 1. Stimulus parameters

Primaries		Ratio f_2/f_1	Geometric mean of f_1 and f_2, kHz	DP $2f_1-f_2$ Hz
f_1 (73 dB) Hz	f_2 (67 dB) Hz			
430	581	1.35	0.5	280
877	1,140	1.3	1	614
1,328	1,694	1.275	1.5	963
1,789	2,236	1.25	2	1,342
2,710	3,320	1.225	3	2,101
3,652	4,382	1.2	4	2,921
5,535	6,504	1.175	6	4,567
7,460	8,579	1.15	8	6,341

acoustic probe sealed into the ear canal. There, the stimuli were mixed acoustically. The levels of the tones were fixed at 73 dB HL for the lower-frequency stimulus (f_1) and at 67 dB HL for the higher-frequency stimulus (f_2). The levels were calibrated in a standard 2-cm^3 coupler.

Eight pairs of stimuli were tested. The frequencies of the two primaries were chosen so that their geometric means represented the standard audiometric frequencies between 0.5 and 8 kHz. The ratios f_2/f_1 were based on the results reported by Harris et al. [8]. The parameters of the stimuli are summarized in table 1. They were fixed throughout the experiment.

It was assumed that the DPs were generated at the geometric mean frequency of the two pure-tone stimuli, and that, therefore, they represented this frequency. This concept is based on many experimental findings indicating that the DP $2f_1-f_2$ is generated at the frequency region of the primaries and not at the frequency of the DP itself [9–11]. The geometric mean frequency of the primaries was recommended as the best single frequency to which the DP $2f_1-f_2$ generation should be related [12, 13].

Acoustic Probe and Measurement of DPOAE

The DPOAEs were recorded by a calibrated ear-canal probe (Etymotic Research ER-10) containing two ports for the insert earphones and a sensitive, low-noise microphone system. The probe was designed to provide relatively flat frequency responses for the stimuli and for the microphone system within the human ear canal. An associated low-noise preamplifier equalized the frequency response of the microphone system additionaly. The probe was sealed into the ear canal by a soft sponge tip.

After preamplification the signal was delivered to a dynamic-signal analyzer (Hewlett-Packard 3561A). Measurements of the DPOAE amplitudes at $2f_1-f_2$ and the adjacent noise floor were taken by averaging eight spectra. The frequency analysis was centered at the DPOAE frequency, the frequency spans were 200, 400 or 500 Hz. The primaries were not included in the windows. Using a Hanning weighting function, the corresponding analysis bandwidth were 0.75, 1.5 and 1.875 Hz. The input range was set to a fixed value at quiet conditions before the actual measurement, and an artifact rejection system accepted only averages that did not exceed this value.

Emissions were considered to be present if the DPOAE amplitude was at least 6 dB above the noise floor. The value of the noise floor was measured by taking a visual mean of its random deflections within the averaged FFTs. The DPOAE amplitudes were always measured at the fixed frequencies of $2f_1-f_2$. The level of each DPOAE frequency was calibrated in a 2-cm^3 coupler.

The absence of DPs above the noise floor at $2f_1-f_2$ in the stimulating and recording system was confirmed in various cavities of volumes between 0.5 and 2 cm^3.

Results

The noise floor at the DPOAE frequencies during stimulation was averaged for the three groups of subjects (normals, near-normals, patients) separately. The mean noise floor varied from − 17 to 8 dB SPL, depending on frequency and group. The standard deviations (SDs) were between 3 and 8 dB SPL. The noise floors were essentially the same for the three groups with the exception of the lowest frequency which was higher in the patient group (normals: 2.5 ± 5.3 dB; near-normals 3.9 ± 4.2 dB; patients 8 ± 7.4 dB). The noise floors for the three groups are shown in figure 1–3.

The amplitudes of DPOAEs in the normal group ranged from 0 to 40 dB above the noise floor. The mean amplitudes depended on the frequency. This is depicted in figure 1. With the exceptions of the lowest and highest frequency, all mean DPOAEs were more than 10 dB above the noise floor. The SDs were around 7 dB.

Similar findings, shown in figure 2, were obtained in the near-normal group. The mean DPOAE amplitudes exceeded 10 dB above the noise floor in the frequency range between 1 and 4 kHz. This value was 9 dB at 6 kHz, compared to 10.8 dB in the normal group. Therefore, normal and near-normal subjects were pooled together for further analysis. If a DPOAE amplitude of 6 dB above the noise floor is accepted as the lowest limit for DPOAE detection, 85–98% of all normal and near-normal ears showed DPOAEs when each frequency between 1 and 6 kHz was considered separately. Emissions in three of these 6 frequencies were detected in all but 2 of the 113 ears (98.2%). However, it was relatively common that the DPOAE at a single frequency was less than 6 dB above the noise floor. Fifty-four percent of these ears had DPOAEs of 6 dB above the noise floor at all frequencies between 1 and 6 kHz. At 0.5 and 8 kHz, this percentage was only 16 and 33%, respectively.

Figure 3 shows the averaged results in the patient group. The mean DPOAE amplitudes exceeded 10 dB above the noise floor at only 1 and 1.5 kHz. However, the amplitudes and the frequency distribution of DPOAEs were dependent on the amount and shape of audiometric hearing losses in pathological ears. These losses were of widely differing degrees. Analysis at single frequencies was carried out. An example for such an analysis at 1 kHz is depicted in figure 4. The hearing thresholds of all normal and pathological ears (n = 199) are plotted in this figure against their DPOAE amplitudes. Though there is a clear tendency of lower DPOAE amplitudes at higher hearing thresholds, the spread is considerable. Inspection of figure 4 reveals that some ears with an audiometric hearing loss of as much as 65 dB still showed DPOAEs with amplitudes around 10 dB. On the other hand, several ears with normal hearing did not generate DPOAEs of more than 6 dB at this frequency.

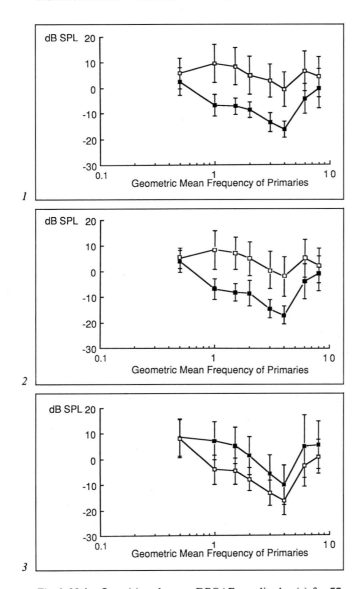

Fig. 1. Noise floor (■) and mean DPOAE amplitudes (□) for 77 normally hearing ears, related to the geometric mean frequency of the primaries. Vertical bars indicate ± 1 SD.

Fig. 2. Noise floor (■) and mean DPOAE amplitudes (□) for 36 near-normally hearing ears. See also legend of figure 1.

Fig. 3. Noise floor (■) and mean DPOAE-amplitudes (□) for pathological ears. See also legend of figure 1.

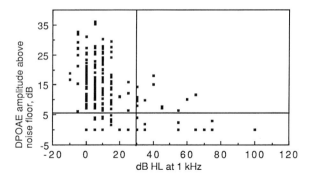

Fig. 4. DPOAE amplitudes at 1 kHz geometric mean frequency of the primaries, related to the audiometric hearing threshold at 1 kHz. Each point represents one ear (n = 199). The horizontal line marks the 6-dB limit of DPOAE amplitudes, the vertical line indicates a hearing loss of 30 dB as an arbitrary limit between normal and pathological hearing.

Discussion

Emissions of DPs may have important clinical applications. For clinical purposes, the pure-tone audiogram with its sampled frequencies tested over the most important range of hearing has proven to be highly valuable for diagnostic and therapeutic decisions. Transiently evoked OAEs (TEOAEs) can already be considered as a tool for objective hearing evaluation [14–16]. When compared to the pure-tone audiogram, a difficulty may arise with the use of TEOAEs: they are often dominated by certain frequencies that cannot be chosen by the examiner. The number and values of these frequencies are highly variable from ear to ear [15].

Distortion product emissions, on the other hand, can be evoked at chosen frequencies, depending on the frequencies of the primaries. This study presented evidence that DPOAEs can be recorded in a clinical setting using relatively simple methods. However, several unsolved problems remain, including the parameters to select, the verification of DPOAEs, the testable frequency range, the frequency specificity, and the variability of individual ears.

The multitude of parameters that have to be considered and that can be varied when testing DPOAEs represent a primary difficulty. Examples are the frequency ratio of the primaries, their level ratio, the particular DPs to be measured, and the frequency regions where the primaries, the DP frequency and the DP generation site are located. A relatively simple set of parameters was used in this study. Obviously, many other combinations are possible. Only systematic studies of these parameters in clinical populations, such as the one reported by Harris et al. [8], will reveal which parameters are ideal for what purposes.

The low level of human DPOAEs that were found to be 70 dB smaller than the primaries in this study present an additional difficulty. This value is in good agreement with other studies [8, 13, 17]. At these low levels, many sound systems develop DPs related to the equipment. Therefore, a verification of biologically produced DP is necessary. Very effective verification procedures are generally used in the clinical applications of TEOAEs [14, 15, 18]. Similar procedures are not yet available for DPOAEs. In this study, the verification rested on the demonstration that the measuring system did not produce DPs in dead ears or in small couplers. Other procedures may be helpful, for instance input-output functions of DPOAEs at each frequency examined. Biologically produced DPs tend to show some saturation at medium to high stimulus levels [13], whereas artifactual DPs tend to grow exponentially. Suppression of the DPOAE by a third tone may be another possibility of verification.

The frequency of the DPOAE generation site was related to the geometric mean of the primaries [9–13] and the useful frequency range was found to lie between 1 and 6 kHz. An extension of this range would be clinically helpful, especially to lower frequencies. Possibly, improved techniques for the detection of signals in noise would permit this range extension. Since the measurement for 500 Hz was taken at 280 Hz, biological and system background noise probably interfered with the detection of DPOAEs at such low frequencies. The noise floor up to about 1 kHz was probably contaminated by such noise sources in our system (fig. 1–3). The rise at higher frequencies, however, was more likely related to our particular choice of parameters.

A good separation between the noise floor and the DPOAE amplitudes was obtained for the frequency range 1–6 kHz in normals and near-normals. All findings were similar when these two groups were compared. Thus, mild and symptomless hearing loss does not seem to change the basic properties of the DPOAEs. Normal hearing can be defined quite broadly with respect to DPOAEs.

Only about half of these normal ears showed DPOAEs at all six frequencies tested between 1–6 kHz. Several explanations can be provided for the absence of DPOAEs at single frequencies in normally hearing ears. The amplitude related to the noise floor could be small, either because the

DPOAE was smaller for the particular ear at this specific frequency, or because the noise floor was temporarily higher due to reasons such as head movements or swallowing. Individual variations of the middle ear mechanics may lower the DPOAEs at specific frequencies [19]. Moreover, the relationship between the pure tone audiogram and the assumed DPOAE-generation site at the geometric mean frequency of the primaries is an approximation. The same frequency specificity of DPOAE-testing as in the pure tone audiogram should not be expected. Several frequencies are involved in DPOAE-testing and the frequencies of the primaries cannot be too close together [9]. Additionally, the stimuli are presented at levels well above threshold causing a spread of energy along the basilar membrane.

Similarly, the relationship between DPOAE amplitudes and hearing thresholds cannot be described in simple terms for both normals and patients. This fact is demonstrated in figure 4. In patient ears, the overall individual picture often seemed to indicate the audiometric shape of hearing loss better than single frequencies, but further analysis of the different pathologies and shapes of hearing losses is necessary. A limit of 6 dB above the noise floor was assumed for a separation between the presence or absence of DPOAEs in this study. As shown in figure 4, this limit is quite effective but far from perfect. Further basic and clinical research should try to explain the variability of DPOAEs in human ears, as well as to establish their relationships to other measures of auditory functions.

In summary, we have demonstrated that the measurement of DPOAEs is feasible in a clinical setting. Additionally, there is a significant relationship between the DPOAE amplitude and the hearing threshold at specific, chosen frequencies. Therefore, some degree of frequency specificity can be obtained. These results, and those of others [13, 17], are promising for a future clinical use of DPOAEs. However, further investigations are necessary before DPOAEs can be accepted as a routine clinical tool. Analyses of the influence of the middle ear function and the relationship of DPOAEs to certain pathologies are only two of many additional areas which have to be studied further.

References

1   Helmholtz H: Die Lehre von den Tonempfindungen, als physiologische Grundlage für die Theorie der Musik, ed 3. Braunschweig, Vieweg, 1870.
2   Békésy G von: Über die nichtlinearen Verzerrungen des Ohres. Ann Phys Lpz 1934; 20:809–811.

3   Zwicker E: Der ungewöhnliche Amplitudengang der nichtlinearen Verzerrung des
     Ohres. Acustica 1955;5:67–74.
4   Plomp R: Detectability thresholds for combination tones. J Acoust Soc Am 1965;37:
     1110–1123.
5   Goldstein JL: Auditory nonlinearity. J Acoust Soc Am 1967;41:676–689.
6   Kemp DT: Stimulated acoustic emissions from within the human auditory system. J
     Acoust Soc Am 1978;64:1386–1391.
7   Kemp DT: Evidence of mechanical nonlinearity and frequency selective wave
     amplification in the cochlea. Arch Otorhinolaryngol 1979;224:37–45.
8   Harris FP, Lonsbury-Martin BL, Stagner BB, et al: Acoustic distortion products in
     humans: Systematic changes in amplitudes as a function of f_2/f_1 ratio. J Acoust Soc
     Am 1989;85:220–229.
9   Kemp DT: Otoacoustic emissions, travelling waves and cochlear mechanisms. Hear
     Res 1986;22:95–104.
10   Zwicker E: Suppression and $(2f_1-f_2)$ difference tones in a nonlinear cochlear prepro-
     cessing model with active feedback. J Acoust Soc Am 1986;80:163–176.
11   Martin GK, Lonsbury-Martin BL, Probst R, et al: Acoustic distortion products in
     rabbit ear canal. II. Sites of origin revealed by suppression contours and pure-tone
     exposures. Hear Res 1987;28:191–208.
12   Schmiedt RA: Acoustic distortion in the ear canal. I. Cubic difference tones: Effects
     of acute noise injury. J Acoust Soc Am 1986;79:1481–1490.
13   Lonsbury-Martin BL, Harris FP, Hawkins MD, et al: Acoustic distortion products in
     humans. Abstr 11th Midwinter Res Meet Ass Res Otolaryngol, St. Petersburg Beach,
     1988, p 178.
14   Kemp DT, Bray P, Alexander L, et al: Acoustic emission cochleography – Practical
     aspects. Scand Audiol 1986;25(suppl):71–82.
15   Probst R, Lonsbury-Martin BL, Martin GK, et al: Otoacoustic emissions in ears with
     hearing loss. Am J Otolaryngol 1987;8:73–81.
16   Bonfils P, Uziel A, Pujol R: Evoked oto-acoustic emissions from adults and infants:
     clinical applications. Acta Otolaryngol 1988;105:445–449.
17   Harris FP, Glattke TJ: Distortion-product emissions in human with high-frequency
     sensorineural hearing loss. J Acoust Soc Am 1988;84(suppl 1):S74.
18   Probst R, Coats AC, Martin GK, et al: Spontaneous, click-, and tone burst-evoked
     otoacoustic emissions from normal ears. Hear Res 1986;21:261–275.
19   Matthews JW, Molnar CE: Modeling intracochlear and ear canal distortion product
     $(2f_1-f_2)$; in Allen JB, Hall JL, Hubbard A, et al (eds): Peripheral Auditory Mecha-
     nisms. Berlin, Springer, 1986, pp 258–265.

PD Dr. R. Probst, HNO-Universitätsklinik, CH–4031 Basel (Switzerland)

Grandori F, Cianfrone G, Kemp DT (eds): Cochlear Mechanisms and
Otoacoustic Emissions. Adv Audiol. Basel, Karger, 1990, vol 7, pp 126–138

Distortion Product Otoacoustic Emissions and Spontaneous Otoacoustic Emission Suppression in Humans

G. Cianfrone[a], *M. Mattia*[a, b], *G. Altissimi*[a], *R. Turchetta*[a]

[a] Auditory Research Laboratory, II ENT Department, University 'La Sapienza',
and [b] Brüel & Kjaer, Research Center, Rome, Italy

Distortion products (DPs), known in the past as 'combination tones' –
term introduced by Vieth in 1805 – can be of different orders [1, 2],
depending on the combination between two external tones, f_1 and f_2. The
third-order combination tones correspond to the combination $(2f_1–f_2)$,
where f_1 is lower in frequency than f_2. These tones are called 'cubic difference tones' (CDTs) and are the most extensively investigated and used
both because they have the highest intensity, thus being more easily generated, and because it is believed that they are the most clearly linked to
cochlear active and nonlinear mechanisms [1–5].

DPs give rise to many questions, especially as far as their sites and
mechanisms of generation are concerned, namely: does the $(2f_1–f_2)$ DP
result from the mechanical interference of two waves generated on the
basilar membrane in the frequency sites of the primary tones, f_1 and f_2, or
is it produced by one wave generated in the CDT frequency site on the
basilar membrane? According to the latter hypothesis, the DP might derive
from one source alone and from an active mechanism located just in that
region. Based on current scientific knowledge, it cannot be ruled out that
the two mechanisms – interaction of two waves and active generation of
one wave – may operate simultaneously [11].

The clinical interest that is presently focused upon DPs is mainly due
to: the possibility of their measure by the use of the otacoustic emission
(OAEs) technique, that is to say in an objective, noninvasive way
(DPOAEs); and, which is more important from the clinical standpoint, the
possibility of performing all measurements by testing also human ears.

Furthermore, CDTs are characterized by the following features: (1) they can be universally evoked in all normally hearing subjects; (2) they show very good test-retest reproducibility; (3) they allow testing of different areas of the cochlear partition, by appropriately varying the two external primary tones, f_1 and f_2; (4) they show an early vulnerability to stimuli harmful to the organ of Corti (noise exposure, ototoxic drugs, endolymphatic hydrops, etc.) [16–19].

It should be taken into account that 30–40% of normally hearing subjects, according to available normative data [20, 21] show one or more spontaneous OAEs (SOAEs) of sizeable intensity, scattered within frequency ranges which turn out to be of great practical, and especially diagnostic interest (i.e., from 1,000 to 4,000 Hz). Accordingly, it appeared essential to test the following: (a) whether the mechanism of generation and the parameters of the CDT are actually affected by the presence of SOAEs near the DP, as pointed out by some authors [8, 12, 22, 23]; (b) whether the suppression of an SOAE may be influenced by the presence and growth of a CDT, independently of or in association with the effect of the external suppressor tones [24]; (c) whether the reciprocal frequency position of SOAEs and CDT (apical or basal i.e. lower or higher) may influence the interactions between them.

All these possible interferences should be clearly known and evaluated in the measurement of DPs in the normal ears in order not to misinterpret data from hearing-impaired subjects [25–27].

Method

The tests were performed on a 22-year-old girl, without hearing impairment which could be tested by pure tone audiometry, multifrequency audiometry at intervals of 65 Hz, and impedance audiometry. Her clinical history and otoscopy were negative. In this subject, bilateral SOAEs – multiple and unstable on the left side, single and stable on the right – had already been detected. Thus, the right side was chosen for our investigation, where the SOAE was of 2,288 Hz with a 6- to 7-dB SPL (fig. 1).

The equipment used to carry out our experiments included:

For the analysis of SOAEs and DPOAEs: (1) a ½-inch condenser microphone (Brüel and Kjaer 4166), characterized by a flat and linear response from 2.6 Hz to 10 kHz (± 2 dB, introduced into a special conical device coupled, at the furthermost end, to the external ear canal; (2) a low-noise preamplifier (Brüel and Kjaer 2660); (3) a two-channel FFT analyzer (Brüel and Kjaer 2032). The recording system (microphone, preamplifier and FFT analyzer) enabled us to reduce the system noise floor to -15 dB SPL with a resolution of 4 Hz. This chain can undergo up to 146 dB SPL with a limit of 3% distortion; (4) a Hewlett-Packard 9816 personal computer for data processing and handling.

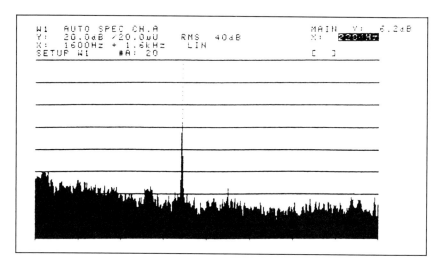

Fig. 1. The configuration of the SOAE on the right ear chosen for our measurements (2,288 Hz, 6.2 dB SPL).

For the generation of the two primary tones, f_1 and f_2: two independent sine generators (Starkey HSL Laboratory), and 2 separate transducers, introduced in the coupling device described above, which also contained the recording microphone.

To calibrate the measuring system, we used a Brüel and Kjaer 4220 pistonphone with a nominal 124 ± 0.2 dB SPL at 250 Hz. As to the possible instrumental distortion, it was tested before each recording session. The results of the tests showed that it was possible to operate safely up to 75–85 dB SPL, depending on the frequency of the primaries.

The primary tones, f_1 and f_2, have been selected so as to produce a DPOAE just apical (lower frequency) or just basal (higher frequency) to the SOAE at four different frequency separations between CDT and SOAE: 28 and 208 Hz with the CDT basal to the SOAE, 78 and 180 Hz, with the CDT apical to the SOAE. The f_1/f_2 frequency ratio ranged from 0.85 to 0.92.

The measurements have been performed during two separate and subsequent sessions. During the first sessions, the investigation was focused on the suppression of the SOAE and the generation of the CDT, using two external tones selected so as to have: (a) the CDT lower in frequency than (apical to) the SOAE with a Δf of 78 Hz between SOAE and CDT; (b) the CDT higher in frequency than (basal to) the SOAE with a 28 Hz Δf between SOAE and CDT. During the second session, performed a few days later, the same procedure was used, but changing the Δf between SOAE and CDT as follows: (a) CDT lower than SOAE with a Δf of 180 Hz; (b) CDT higher than SOAE with Δf of 208 Hz.

In each measurement: (1) f_1 and f_2 were kept equal in level; (2) an ascending approach was used, that is starting from lower primary levels to continue with increas-

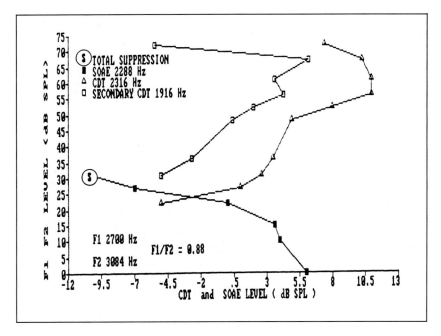

Fig. 7. Input-output amplitude function for SOAE and DPOAEs (main and secondary). Note the relative parallelism between the main and the secondary CDT.

Discussion

The results of the tests may be interpreted according to different keys: (1) general effects of the presence of an SOAE on a CDT close to it; (2) relation between the suppression of the SOAE and the generation of the CDT; (3) different effects of the SOAE on the CDT, depending on their mutual position in terms of frequency (i.e. CDT basal or apical to the SOAE).

(1) Kemp [8] had already formulated the hypothesis that in the cochlear mechanical system, the nonlinearity, which is expressed by the presence of OAEs, enhances the generation of harmonic or intermodulation distortions. In 1988, Wier et al. [18], referring to this hypothesis, tried to prove this kind of influence. They actually found out that in a range of about 100 Hz above or below an SOAE, the amplitude of a CDT is 10–25 dB higher than in a situation in which SOAEs are completely absent. Our results confirm these data, since for the smaller frequency separations used

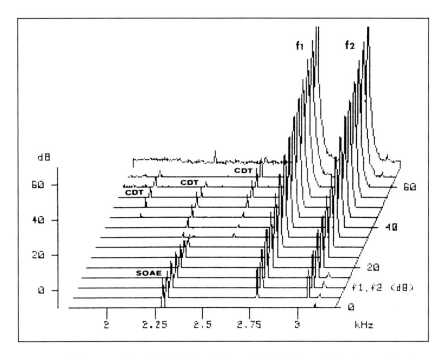

Fig. 8. Two secondary DPOAEs (at 2,228 and 1,960 Hz) are clearly detectable, when the main DPOAE sets in a basal position (i.e. higher frequency) with respect to the SOAE.

between SOAE and CDT (28 and 78 Hz), the minimum level of primaries capable of generating a CDT is 23–25 dB, whereas for the higher separations (180 and 208 Hz), the minimum level is 30–35 dB. The analysis of the growth function of the CDT close to a SOAE is more complex. Some evidence of the influence is provided by the fact that the intensity level of the primaries corresponding to the saturation point of the CDT is different in relation to the different frequency separations between CDT and SOAE. This direct interaction between SOAE and CDT, whose effect is the amplification of the latter produced by the former, might substantiate the assumption formulated by some authors [11] that the DP develops active-ly, at least partially, in the place of the cochlear partition corresponding to the $(2f_1-f_2)$ frequency. Therefore, this effect of amplification or enhance-ment produced on the CDT might be linked, at least partially, to complex interaction phenomena, including synchronization and phase locking, between two self-sustaining active oscillators (SOAE and CDT), located close to each other [23].

(2) The investigation on the suppression of the SOAE and the generation of the CDT occurring at the same time, enable us to check the reliability of the hypothesis of some authors [18] that, when both SOAEs and DPs exist, if they are close enough, the major source of suppression of the SOAE is the generation of the CDT, not the external stimulation f_1 or f_2. A direct action of the CDT on the suppression of the SOAE appears to us to be unlikely, since in our measurements the SOAE is almost completely suppressed when the CDT has not yet developed or is around the threshold.

(3) Attention was focussed on the different behaviors of distourtion phenomena, produced by f_1 and f_2, depending on whether the CDT was located apicalward (lower frequency) or basalward (higher frequency) of the SOAE. This difference in behaviors, seen in the first session and detected again in the second session, consists, in practice, in the arising of secondary CDTs. Their onset always and only occurred when f_1 and f_2 had been adjusted so that the primary CDT was basal (higher frequency) to the SOAE, that is between the SOAE and the primary tones; whereas when the main CDT was apical (lower frequency) to the SOAE, no other distortion component could be observed. During the first session, it was possible to detect one secondary distortion product, whose intensity was quite high. Conversely, during the second session, slightly changing the f_1 and f_2 values, two or more satellite components could be identified. All these secondary DPs may result from different third-order or higher combinations between f_1 and f_2, between f_1 and the main CDT or between f_2 and the main CDT; the SOAE is not involved in these combinations.

In the different test conditions we have sometimes noticed moderate frequency shifts of the SOAE no larger than 10 Hz without connections with particular and specific experimental conditions. Therefore we do not consider these data to be of particular significance.

The interpretation of such a complex set of data concerning the 'jumping effect' of the CDT over the SOAE is quite difficult. The importance of the apical or basal location of the SOAE to the CDT in the generation of secondary DPOAEs seems to be linked to the different kind and degree of mechanical interference between the propagation waves of the SOAE and the CDT along the basilar membrane under the two different conditions, considering also the bidirectional nature of each wave produced by the SOAE and the CDT and, accordingly, the two-fold possibility of mechanical interference [28].

In conclusion: (a) the quite complex interaction and interference between SOAEs and DPOAEs is well documented; (b) these interferences are strongly different depending on both the frequency separation between the SOAE and the DPOAE and on whether the DP is located apicalward or basalward of the SOAE.

We want finally to point out an intriguing question: may the SOAEs and the DPOAEs be compared to external tones with the same frequency and level? An SOAE, even very strong, is generally not perceived by the emitting subject [29], while an external tone with equal physical characteristics is clearly heard by the same ear. The CDT, even weak, is usually heard and distinguished from the primary tones, while an external tone equal in frequency and level to the CDT and delivered simultaneously with added tones, equal in frequency and level to f_1 and f_2, is not perceived because it is masked and cancelled by the added external tones (f_1-like, f_2-like tone, or both).

In our measurements we have shown otoacoustically that the DPOAE arises and develops just inside the suppression tuning curve of f_1, the proof is that the SOAE lying in close vicinity is completely suppressed by f_1.

Our explanation is the following: the SOAEs, whose origin is likely directly related to the efferent system (outer hair cells), are unable to pass into the afferent sensorial system because identified as a biological 'nuisance' and thus are inhibited by the same efferent control system. DPs arising from efferent or afferent elements do not undergo an inhibitory action but rather a facilitation effect because they are identified as 'useful' signals. So it is possible to suggest that SOAEs and CDTs may originate from different structures or, even if from the same structures, the influence exerted by the efferent control system may be of opposite sign.

References

1   Plomp R: Detectability threshold for combination tones. J Acoust Soc Am 1965;37: 1110–1123.
2   Smoorenburg GF: Combination tones and their origin. J Acoust Soc Am 1972;52: 615–631.
3   Zwicker E: Der ungewöhnliche Amplitudengang der nichtlinearen Verzerrungen des Ohres. Acustica 1955;5:67–74.
4   Zwicker E: Suppression and ($2f_1–f_2$) difference tones in a nonlinear cochlear preprocessing model with active feedback. J Acoust Soc Am 1986;80:163–176.
5   Goldstein JR: Auditory nonlinearity. J Acoust Soc Am 1967;41:676–689.

6   Dallos P: Combination tone $2f_1-f_2$ in microphonic potentials. J Acoust Soc Am 1969;46:1437-1444.
7   Hall JL: Two-tone distortion products in a nonlinear model of the basilar membrane. J Acoust Soc Am 1974;56:1818-1827.
8   Kemp DT: Evidence of mechanical nonlinearity and frequency selective wave amplification in the cochlea. Arch Otorhinolaryngol 1979;224:37-45.
9   Formby C, Sachs RM: Psychophysical tuning curves for combination tones $2f_1-f_2$ and f_2-f_1. J Acoust Soc Am 1980;67:1754-1758.
10 Kim DO: Cochlear mechanics: Implications of electrophysiological and acoustical observations. Hear Res 1980;2:297-317.
11 Kim DO, Molnar CE, Matthews JW: Cochlear mechanics: Nonlinear behavior in two-tones responses as reflected in cochlear-nerve-fiber responses and in ear-canal sound pressure. J Acoust Soc Am 1980;67:1704-1719.
12 Burns EM, Strickland EA, Tubis A, Jones K: Interactions among spontaneous otoacoustic emissions. I. Distortion products and linked emissions. Hear Res 1984;16:271-278.
13 Kim DO: Active and nonlinear cochlear biomechanics and the role of outer-hair-cell subsystem in the mammalian auditory system. Hear Res 1986;22:105-114.
14 Strube HW: The shape of the nonlinearity generating the combination tone $2f_1-f_2$. J Acoust Soc Am 1986;79:1511-1518.
15 Martin GK, Lonsbury-Martin BL, Probst R, Scheinin SA, Coats AC: Acoustic products in rabbit ear canal. II. Sites of origin revealed by suppression contours and pure-tone exposure. Hear Res 1987;28:191-208.
16 Siegel JH, Kim DO: Cochlear biomechanics: Vulnerability to acoustic trauma and other alterations as seen in neural responses and ear-canal sound pressure; in Hamernik RP, Henderson D, Salvi R: New Perspectives on Noise-Induced Hearing Loss. New York, Raven Press, 1982, pp 137-151.
17 Norton SJ: Effects of auditory fatigue on psychophysical estimates of cochlear nonlinearities. J Acoust Soc Am 1987;82:80-87.
18 Wier CC, Pasanen EC, Mc Fadden D: Partial dissociation of spontaneous otoacoustic emissions and distortion products during aspirin use in humans. J Acoust Soc Am 1988;84:230-237.
19 Horner KC, Cazals Y: Distortion associated with endolymphatic hydrops in the guinea pig. Valsalva. In press. 1989.
20 Zurek E: Spontaneous narrow-band acoustic signals emitted by human ears. J Acoust Soc Am 1981;69:514-523.
21 Cianfrone G, Mattia M: Spontaneous otoacoustic emissions from normal human ears. Preliminary report. Scand Audiol 1986;suppl 25:121-127.
22 Frick LR, Matthies ML: Effects of external stimuli on spontaneous otoacoustic emissions. Ear Hear 1988;9:190-197.
23 van Dijk P, Wit HP: Phase-lock of spontaneous otoacoustic emissions to a cubic difference tones; in Basic Issues in Hearing. London, Academic Press, 1988, pp 101-105.
24 Cianfrone G, Mattia M: Spontaneous otoacoustic emissions: Suppression induced by external tones; in Feldmann H (ed): Proc III Int Tinnitus Seminar. Karlsruhe, Harsch, 1987, pp 84-86.

25 Harris FP: Distortion product emissions and pure tone behavioral thresholds. Valsalva. In press, 1989.
26 Leonard G, Smurzynski J, Jung M, Kim DO: Evaluation of distortion product otoacoustic emissions as a basis for the objective clinical assessment of cochlear function. Proc 2nd Int Symp Cochlear Mechanics and Otoacoustic Emissions, Rome 1989. Adv Audiol. Basel, Karger, 1990, vol 7, pp 139–148.
27 Probst R, Antonelli C, Pieren C: Methods and preliminary results of measurements of distortion product otoacoustic emissions in normal and pathological ears. Proc 2nd Int Symp Cochlear Mechanics and Otoacoustic Emissions, Rome 1989. Adv. Audiol. Basel, Karger, 1990, vol 7, pp 117–125.
28 Kemp DT: Otoacoustic emissions, travelling waves and cochlear mechanisms. Hear Res 1986;22:95–104.
29 Wilson JP, Sutton GJ: A family with high-tonal objective tinnitus. An update; in: Klinke R, Hartmann R, Hearing-physiological Bases and Psychophysics. Berlin, Springer, 1983, pp 97–103.

G. Cianfrone, Auditory Research Laboratory, II ENT Department,
University 'La Sapienza', I–00161 Rome (Italy)

Grandori F, Cianfrone G, Kemp DT (eds): Cochlear Mechanisms and
Otoacoustic Emissions. Adv Audiol. Basel, Karger, 1990, vol 7, pp 139–148

Evaluation of Distortion Product Otoacoustic Emissions as a Basis for the Objective Clinical Assessment of Cochlear Function

G. Leonard[a], *J. Smurzynski*[a], *M.D. Jung*[a], *D.O. Kim*[a, b]

[a] Division of Otolaryngology, Department of Surgery,
University of Connecticut Health Center, and
[b] Neuroscience Program, University of Connecticut Health Center,
Farmington, Conn., USA

Otoacoustic emissions, both spontaneous and evoked of various types, have been described in the literature in recent years [1–3]. Though these studies suggest their possible clinical application in the assessment of cochlear function, we believe their measurement and evaluation as a clinical tool remains to be accomplished. In our opinion, at the present time, spontaneous emissions cannot be used as a reliable indicator of cochlear function, since they occur only in about 30% of normally hearing human ears, and are difficult to correlate with the functional state of the cochlea.

It is most likely that all evoked emissions – click-evoked, tone-pip evoked, and distortion product emissions (DPOAEs) – will have clinical application under different clinical circumstances. Distortion product emissions are generated by normal ears in response to two tonal stimuli, having the primary frequencies f_1 and f_2 ($f_2 > f_1$). These distortion emissions, the most prominent of which occurs at $2f_1-f_2$, are detectable by a microphone placed in the external auditory canal [e.g. 2, 4]. Since presently available data support the view that the DPOAEs are reduced or abolished by impairment of cochlear function in the region of f_1 and f_2 [2, 5], it is reasonable to suggest that the DPOAEs may be affected in the presence of a hearing loss in the region of f_1 and f_2. There is limited information in the literature about DPOAEs in human ears at the present time

[6–8] (see also Probst et al., present volume). Consequently, we decided to investigate its possible use as a clinical tool.

Our project was designed (1) to demonstrate the incidence and the characteristics, both frequency and level, of DPOAEs in normal human ears, and (2) to demonstrate any correlation between the characteristics of DPOAEs and hearing losses of differing configurations.

Method and Subjects

Method

A probe placed in the external auditory canal, containing a microphone (Knowles EA 1842) and two earphones (Knowles 1716), was used to measure the DPOAEs. Two equilevel pure-tone signals, f_1 and f_2 ($f_2 > f_1$) were generated separately, using two oscillators (Hewlett-Packard 239 A), and were attenuated separately, using two attenuator sets (Hewlett-Packard 350 D). The amplitude spectrum of the microphone signal was analyzed by a sweep-frequency wave analyzer (Hewlett-Packard 3580 A), and was displayed on an X-Y recorder (Hewlett-Packard 7035 B). A 30-Hz bandwidth was used to analyze the microphone output of the probe.

Distortion product emissions were measured in response to three pairs of primary tones for the middle-frequency region: (1) $f_1 = 1.7$ kHz, $f_2 = 1.9$ kHz; (2) $f_1 = 1.85$ kHz, $f_2 = 2.2$ kHz; (3) $f_1 = 2.0$ kHz, $f_2 = 2.5$ kHz and three pairs for the high-frequency region: (4) $f_1 = 3.4$ kHz, $f_2 = 3.8$ kHz; (5) $f_1 = 3.7$ kHz, $f_2 = 4.4$ kHz; (6) $f_1 = 4.0$ kHz, $f_2 = 5.0$ kHz.

The distortion product frequency for $2f_1-f_2$ was fixed at 1.5 kHz and 3.0 kHz for the middle- and high-frequency sets, respectively. In both cases the values of f_1 and f_2 corresponded to a f_2/f_1 ratio of 1.12, 1.19, and 1.25. After the desired primary tones, f_1 and f_2 were set, the amplitude of each was adjusted to achieve a sound pressure level (SPL) of 50 dB re: 20 μPa for each tone. A spectrum of the ear canal acoustic signal was then recorded starting at 1.3 kHz for the middle-frequency set, and at 2.8 kHz for the high-frequency set. The sweeps were terminated at 200 Hz above f_2. Further spectra were recorded by increasing the SPL in increments of 10 and 20 dB. A similar procedure was carried out in the opposite ear.

The probe was then reinserted into the initially tested ear, and further recordings were made. Recordings were taken from alternate ears, every 5–7 min until the complete range of primary tone levels was covered at every 3–4 dB, to a maximum primary tone level of 80 dB SPL. Alternating the ears was necessary because Kim's [2] study showed a reduction of the DPOAE amplitude when an ear was exposed to a fatiguing stimulus. The lowest level was determined as being the point at which the DPOAE was below the noise floor of the instruments. When the bandwidth of the spectrum analyzer was 30 Hz, the noise floor was −5 to −2 dB SPL for $2f_1-f_2 = 1.5$ kHz, and 0 to +8 dB SPL for $2f_1-f_2 = 3$ kHz (depending on input sensitivity of the analyzer).

For DPOAE measurements, the subject was seated comfortably in a single-walled, sound-attenuated room (Industrial Acoustics Company 401), and was instructed to remain as quiet as possible during the test periods. A typical session required approximately 1 h, which included brief rest periods.

Subjects

Results were collected from 17 normal-hearing human ears from 9 healthy adults ranging in age from 21 to 41 years of age. The criteria for normal hearing were that the pure-tone thresholds be at 10 dB HL or better for the octave frequencies 250–8,000 Hz and for the inter-octave frequencies 3,000 and 6,000 Hz, and that tympanometry showed normal middle-ear function.

Ten subjects with abnormal hearing were included in the study. One subject was diagnosed as having a conductive hearing loss in one ear based on pure tone audiometry, tympanometry, and abnormal acoustic reflexes. Nine subjects (18 ears) were diagnosed as having a sensorineural hearing loss. None of them exhibited signs of retrocochlear involvement based on some of the following differential auditory tests: performance-intensity function for phonetically balanced words (PI-PB), acoustic reflex thresholds, and acoustic reflex decay [9]. It was therefore concluded that the sensorineural hearing loss in each of the ears was cochlear in origin.

Results

The results for four pairs of primary tones are discussed. The detection threshold of the $2f_1–f_2$ DPOAE was determined to be the lowest level of primary tones which produced the $2f_1–f_2$ DPOAE above the noise floor. The hearing threshold was estimated by interpolating the pure-tone audiogram data at the frequency corresponding to the geometric mean of f_1 and f_2.

In figure 1, we plotted the detection threshold of $2f_1–f_2$ DPOAE (dB SPL) for given primary tones in a given ear against the hearing threshold (dB HL) at the mean primary frequency. The four panels of figure 1 correspond to four primary frequency pairs. Arrows represent ears in which the DPOAE could not be detected above the noise floor for primary tone levels up to 80 dB SPL.

For $f_1 = 1.85$ kHz, $f_2 = 2.2$ kHz (fig. 1a), the detection threshold of the DPOAE varied from 38 to 58 dB SPL of primary tones in normal ears, with one exception. This exception was for an ear, with high levels of spontaneous otoacoustic emissions (SOAEs) in the vicinity of f_1, f_2, and $2f_1–f_2$, which demonstrated a high DPOAE starting at 32 dB SPL of primary tones. The data points gathered from ears with sensorineural hearing loss suggest an increase in the DPOAE detection threshold of 3–4 dB/10 dB increase in hearing threshold in the 10- to 50-dB HL region.

For $f_1 = 2$ kHz, $f_2 = 2.5$ kHz (fig. 1b), the DPOAE detection thresholds for normal ears varied between 37 and 56 dB SPL of the primary tones. This range included the ear with the previously mentioned SOAE.

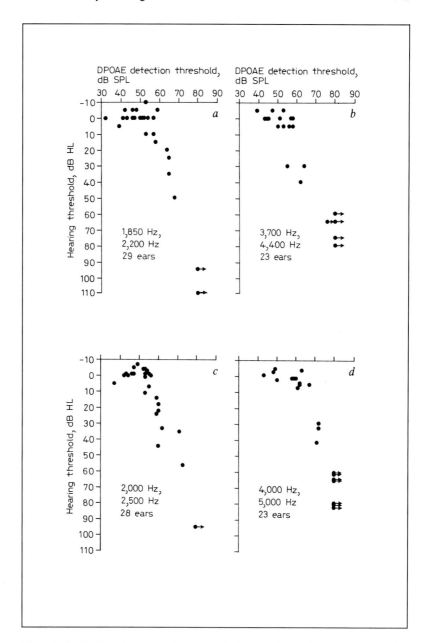

Fig. 1. DPOAE detection thresholds versus the hearing thresholds for four pairs of f_1 and f_2. Arrows indicate ears where the amplitude of the DPOAE was below the noise floor.

At higher primary tone frequencies, $f_1 = 3.7$ kHz, $f_2 = 4.4$ kHz (fig. 1c), normal ears demonstrated DPOAEs at 39–60 dB SPL. In ears with a mild sensorineural hearing loss, the DPOAE detection threshold increased to 55–64 dB SPL. For subjects with severe impairment (higher than 65 dB HL), there was no detectable DPOAE with primary tones presented up to 80 dB SPL.

With primary tones presented at $f_1 = 4$ kHz, and $f_2 = 5$ kHz (fig. 1d), the detection threshold for DPOAEs in normal ears varied between 43 and 67 dB SPL of the primary tones. In ears with a sensorineural hearing loss of 30–40 dB, the DPOAE detection threshold increased to 72 dB SPL and was not detectable in patients with hearing losses greater than 60 dB.

Figure 2 presents the input-output characteristics, demonstrating the relationship between the level of the DPOAE and the level of the primary tones. The data were interpolated so as to plot the DPOAE level for each 2 dB incremental increase in the primary tone level. Each panel in figure 2 shows our total sample of normal ears and examples of hearing impaired ears.

In figure 2, the data gathered from normal ears were divided into two regions of primary tone level. The higher region corresponded to the primary tone level where a distinct DPOAE peak was detected in the spectrum for all sampled ears. For this region both the mean and the standard deviation were calculated. In the lower regions of primary tone level some ears did not demonstrate the $2f_1-f_2$ DPOAE. Therefore, the mean was calculated only for ears with detectable DPOAEs, and the standard deviation was not computed. For primaries around 2 kHz (fig. 2, a,c) the DPOAE was detectable from all normal ears above 56–58 dB SPL, and the standard deviations were between 2 and 6 dB.

The pure-tone audiograms as measured by conventional behavioral audiometry in a sample of 4 subjects are presented in figure 3. Subject 4 (fig. 3) who shows a 'trough-shaped' pattern on audiometry in the region of 2 kHz, in the left ear, has a DPOAE (fig. 2a), below or near the mean minus standard deviation of 17 normal ears. It is notable that the DPOAE from this ear was not detectable until a higher level of primary tones was presented. There was then a dramatic increase in the DPOAE level for a minimal increase in the level of the primary tones, which brought it close to the normal range. We refer to this finding as a 'recruitment-like phenomenon'. This ear also exhibited a 'recruitment-like phenomenon' for the primary tone pair, 2,000 and 2,500 Hz (fig. 2c).

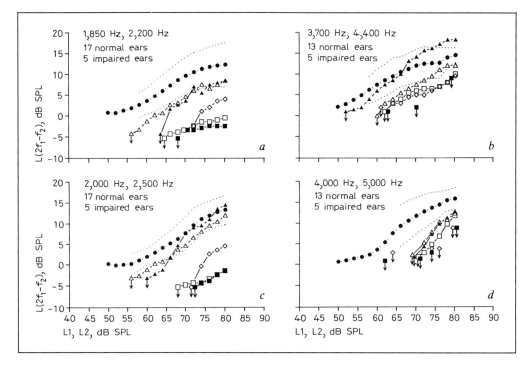

Fig. 2. The relationship between the level of the DPOAE and the level of the primary tones. Mean (●) ± SEM (– – –) for all normal ears. *a, c* Subject 4 (△ = right ear; ▲ = left ear), subject 23 (□ = right ear; ■ = left ear), and subject 14 (◇ = right ear). *b, d* Subject 5 (△ = right ear; ▲ = left ear), subject 13 (□ = right ear; ■ = left ear), and subject 14 (◇ = right ear).

Subject 23 (fig. 3), with a 'ski-slope' pattern audiogram, shows DPOAEs far below those in normal ears for the two middle frequency primary pairs (fig. 2a, c). Furthermore the input-output curves were very shallow.

For primaries f_1 = 3.7 kHz, f_2 = 4.4 kHz, DPOAEs were present in all normal ears above 58 dB SPL of primaries, and at moderate input levels (below 70 dB SPL), the average input-output curve was almost a straight line, with about a 5-dB increase in the DPOAE level for a 10-dB increase in primary tone level. Above 70 dB SPL there was a tendency towards saturation.

Subject 13 (fig. 3), who had a hearing loss of 60 dB at 4 kHz in the left ear, did not demonstrate any DPOAE with primary frequencies near 4 kHz

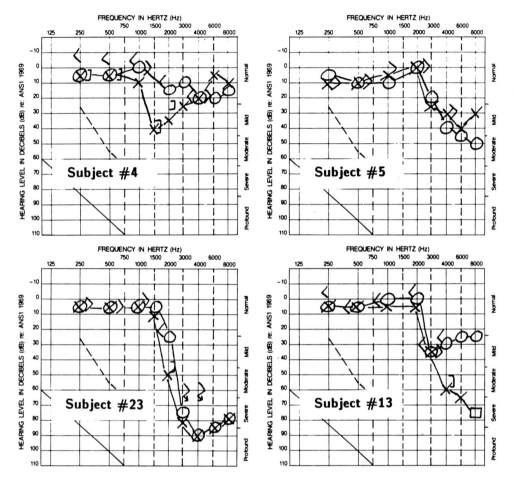

Fig. 3. Pure-tone audiograms from subjects with sensorineural hearing loss. O, X = air conduction for right and left ear; <, > = unmasked bone conduction for right and left ear; [,] = masked bone conduction for right and left ear. Subject 4 left ear: example of 'trough-shaped' pattern; subject 23: example of 'ski-slope' pattern.

(fig. 2b, d). In addition, the right ear, with a 24- to 34-dB hearing loss showed DPOAEs far below the normal range (fig. 2b, d).

An exceptional result was obtained from subject 5 (fig. 3) using primaries 3,700 and 4,400 Hz. The left ear demonstrated a high DPOAE level (particularly for high primary tone levels), even though there was a 30-dB sensorineural hearing loss at 4 kHz in this ear (fig. 2b).

For primaries f_1 = 4 kHz, f_2 = 5 kHz (fig. 2d), some normal ears only demonstrated DPOAEs at rather high primary levels. Subjects 5 and 13 had much smaller DPOAEs than normal ears (fig. 2d).

The right ear of subject 14 (audiogram not shown), with middle ear pathology, showed very small DPOAEs for all primary pairs (fig. 2) and had no emission for primary pair f_1 = 4 kHz, f_2 = 5 kHz.

Discussion and Conclusions

At the outset of this project our goal was to demonstrate (1) the incidence and characteristics of DPOAEs in normal ears, and (2) any correlation that might exist between the characteristics of DPOAEs and hearing losses of differing configurations.

We found that all normal-hearing ears in our sample demonstrated detectable DPOAEs provided that the primary tone level was above a certain value. Our observations are consistent with those of Harris et al. [6]. In contrast, it was reported by Furst et al. [10] that it was difficult to detect DPOAEs in normal human ears which did not have SOAEs. In our study we found several subjects with no SOAEs but relatively high DPOAEs. We also observed that DPOAE was affected by the presence of SOAEs in the vicinity of the primary and DPOAE frequencies. However, we are unable to explain the discrepancy between our data and the data of Furst et al. regarding DPOAEs in ears without SOAEs.

Hearing-impaired ears showed substantially reduced or totally absent DPOAEs. The data points in figure 1 illustrate that ears with a greater sensorineural hearing loss tend to show a higher DPOAE detection threshold. This is further illustrated by subject 13 (fig. 3). The DPOAE was not elicited when the primary frequencies corresponded to the region of severe hearing loss, whereas when the primary frequencies corresponded to the region of normal hearing the DPOAE was within the normal range. Due to the exceptional behavior of some ears (e.g. subject 5), we suggest that it may be necessary to use more than one f_2/f_1 ratio in a frequency region when using DPOAE as a clinical tool.

We found that ears with 'trough-shaped' and 'ski-slope' audiometric patterns exhibited distinct DPOAE input-output characteristics. The ears with a 'ski-slope' pattern exhibited DPOAEs far below the normal range, whereas ears with a 'trough-shaped' pattern exhibited a 'recruitment-like phenomenon'.

The model described by Hall [11] and Kim et al. [12] suggests that the $2f_1-f_2$ DPOAE is generated in the region of the cochlea corresponding to the characteristic place of f_1 and f_2. From that region, the DPOAE is mechanically propagated both to its own characteristic place and to the stapes and, through the middle ear, into the ear canal. We interpret the production of the 'recruitment-like phenomenon' in the following way: (1) an ear with a 'trough-shaped' audiometric pattern is postulated to have a limited abnormal region and a normal basal portion, and (2) this normal basal region allows the generation of the DPOAE at adequately high primary tone levels. With 'ski-slope' pattern ears, the DPOAEs were very small for all levels of primaries. We postulate that the absence of the 'recruitment-like phenomenon' in these ears, with all higher frequencies affected, was due to the absence of a normal basal portion of the cochlea.

Results obtained from the right ear of subject 14 demonstrate that DPOAE detection is affected by abnormalities in the middle ear and it is therefore necessary to rule out middle-ear pathology by acoustic immittance measurements (e.g. tympanometry). Middle-ear pathology is expected to affect the DPOAE in two ways: (1) attenuation of the primary frequency signals going from the earphones to the cochlea, and (2) attenuation of the DPOAE signal going from the cochlea to the ear canal.

Measurements of DPOAEs may provide a noninvasive, objective method of evaluating cochlear function, particularly in certain populations (e.g. infants and difficult-to-test children, the retarded, and malingerers). However, in our opinion the general application of DPOAE measurement as a clinical tool is not yet possible. The data presented in this paper as well as data presented by other studies [6–8] support the view that the DPOAE method, with further refinement and development, will have a place in the measurement of cochlear function.

Acknowledgements

We thank S. Cooper, G. Cote, A. Cymerys, J. McNally, B. Randall, and E. Robinson for serving as volunteer subjects in this study, and H. Nguyen for plotting the data by computer. This study was supported by a grant from the Deafness Research Foundation, a departmental grant from Department of Surgery, University of Connecticut Health Center, and a faculty grant from University of Connecticut Health Center. J. Smurzynski was on leave from Laboratory of Acoustics, Warsaw Academy of Music.

References

1   Kemp DT: Stimulated acoustic emissions from within the human auditory system. J
    Acoust Soc Am 1978;64:1386–1391.
2   Kim DO: Cochlear mechanics: Implications of electrophysiological and acoustical
    observations. Hear Res 1980;2:297–317.
3   Cianfrone G, Grandori F (eds): Cochlear mechanics and otoacoustic emissions.
    Scand Audiol 1986;suppl 25:1–164.
4   Kemp DT: Evidence of mechanical nonlinearity and frequency selective wave
    amplification in the cochlea. Arch Otorhinolaryngol 1979;224:37–45.
5   Zurek PM, Clark WW, Kim DO: The behaviour of acoustic distortion products in
    the ear canals of chinchillas with normal or damaged ears. J Acoust Soc Am 1982;72:
    774–780.
6   Harris FP, Lonsbury-Martin BL, Stagner BB, et al: Acoustic distortion products in
    humans: Systematic changes in amplitude as a function of f_2/f_1 ratio. J Acoust Soc
    Am 1989;85:220–229.
7   Kemp DT, Bray P, Alexander L, et al: Acoustic emission cochleography – Practical
    aspects. Scand Audiol 1986;suppl 25:71–95.
8   Lonsbury-Martin BL, Harris FP, Hawkins MD, et al: Acoustic distortion products in
    humans. Assoc Res Otolaryngol 1988;11:178.
9   Katz J: Handbook of Clinical Audiology. Baltimore, Williams & Wilkins, 1985.
10 Furst M, Rabinowitz WM, Zurek PM: Ear canal acoustic distortion at $2f_1-f_2$ from
    human ears: Relation to other emissions and perceived combination tones. J Acoust
    Soc Am 1988;84:215–221.
11 Hall JL: Two-tone distortion products in a nonlinear model of the basilar mem-
    brane. J Acoust Soc Am 1974;56:1818–1828.
12 Kim DO, Molnar CE, Mattews JW: Cochlear mechanics: Nonlinear behavior in
    two-tone responses as reflected in cochlear-nerve-fiber responses and in ear-canal
    sound pressure. J Acoust Soc Am 1980;67:1704–1721.

Gerald Leonard, MD, Division of Otolaryngology, University of Connecticut
Health Center, Farmington, CT 06032 (USA)

Grandori F, Cianfrone G, Kemp DT (eds): Cochlear Mechanisms and
Otoacoustic Emissions. Adv Audiol. Basel, Karger, 1990, vol 7, pp 149–155

Evoked Otoacoustic Emissions in Sensorineural Hearing Loss: A Clinical Contribution

V. Garrubba[a], *F. Grandori*[b], *M.P. Lamoretti*[a], *A. Antonelli*[a]

[a] Otorhinolaryngologic Clinic of the University, Brescia;
[b] Center of System Theory, Polytechnic of Milan, Italy

In cochlear losses, click evoked otoacoustic emissions (EOAEs) disappear when the average pure-tone threshold is higher than 20–30 dB HL [1–4].

On this basis, Johnsen et al. [5] and Bonfils et al. [6] have employed EOAEs as a method of auditory screening in neonates. On the other hand, Tanaka [6] and Tanaka et al. [7] were able to record EOAEs in subjects with pure-tone averages of 50 dB HL or even more.

In subclinical ototoxic cochlear impairments after the ingestion of loop diuretics [9] or acetylsalicylic acid [10] EOAEs are reduced, thus giving early information of the developing, still reversible, changes occurring in the end organ.

In noise-induced hearing losses, Grandori et al. [11] recorded EOAE in some cases only. In acute auditory fatigue after exposure to noise or pure tones, Kemp [12] showed that EOAEs were clearly affected when the temporary threshold shift was of 5–10 dB, and that recovery paralleled the recovery of the hearing thresholds.

In sensorineural losses, the recording of EOAEs can help in differentiating 'pure' cochlear impairments from lesions involving mainly eighth nerve units. The working hypothesis is that EOAEs should be absent in cochlear cases, whilst they could be normally present when only the eighth nerve is affected.

The type of eighth nerve lesion to be differentiated on the grounds of EOAEs must not give rise to cochlear disorders, for instance, by interfering with its blood supply, like acoustic schwannomas in most cases.

Bonfils et al. [4], out of 28 patients with surgically confirmed acoustic neurinoma, were able to record EOAEs from only 9 of the affected ears, while all the 22 subjects with lesions confined to auditory pathways at the brainstem level produced EOAEs. Our present contribution is aimed at analyzing the presence of EOAEs in some cases with different types of sensorineural hearing losses.

Materials and Methods

Materials
21 normal subjects (42 ears), 11 males, age range 20–35 years, were the control group. The test group consisted of (a) 2 subjects (4 ears) with continuous, bilateral tinnitus, and normal pure-tone threshold and impendance measurements; (b) 2 Ménière's patients (2 ears); (c) 8 subjects (16 ears) with hearing impairment due to chronic noise exposure; (d) 4 subjects (8 ears) with 'familial low-frequency hearing loss' [13].

Methods
Routine examination included an ENT examination, pure-tone thresholds from 0.125 to 8 kHz at octave intervals and impedance measurements. Vestibular testing, speech audiometry and recoding of auditory brainstem responses (ABR) were performed when needed. Patients whose ABR waveform and/or latency of the waves and/or interwave interval were not normal underwent a CT scan to rule out a cerebellopontine angle tumor.

To record EOAEs, the method described by Grandori [14, 15] was used in 11 normal subjects and in the majority of the hearing-impaired subjects (Amplaid MK VII system). In 10 normals, the technique of Bray and Kemp [16] was employed (ILO 88).

Acoustic stimuli were 0.1-ms clicks or 1-kHz tone bursts (rise decay time of 0,5 ms and a plateau duration of 0, 2 or 4 ms) delivered by air conduction. Repetition rate was 21 pps. Intensity level was increased in 10-dB steps from the psychoacoustic threshold for the actual stimulus, to 40 dB above threshold.

Criteria for assessing the presence of EOAEs were: (1) waveform; (2) test-retest reproducibility; (3) spectrum analysis; (4) group latency and latency shifts with change in stimulus level; (5) input/output function of the recorded EOAES; (6) intraindividual longterm stability of the EOAE pattern [17]. In the present contribution, only the presence/absence of the EOAEs will be considered.

Results

In normal subjects, EOAES were detected both with clicks and 1-kHz tone bursts in 39 out of 42 ears (92.8%); this was in agreement with other reports [1, 2, 4, 10, 14, 18] (fig. 1, 2). EOAEs were not recorded in the 2

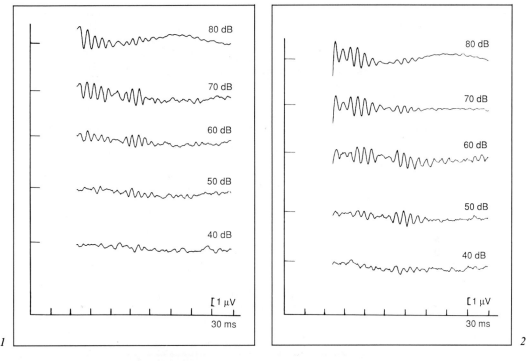

Fig. 1. EOAE to 2-ms 1-kHz tone burst in a normal subject. Method of Grandori [15]. Intensity level in dB SPL.

Fig. 2. EOAE to clicks in a normal subject. Method of Grandori [15]. Intensity level in dB p.e. SPL.

subjects with bilateral tinnitus and in Ménière's ears with pure-tone threshold averages of 45 and 50 dB HL. In the ears with noise-induced losses, EOAEs were usually present when pure-tone threshold at 1 and 2 kHz was lower than 20 dB HL, and absent when the threshold, at the same frequencies, was higher than 30 dB HL (fig. 3, 4). In subjects with 'familial low-frequency hearing loss' two types of sensorineural impairment were found: a 'cochlear' pattern (positive recruitment testing; speech discrimination score of 70% or greater; absence of roll-over in the performance-intensity function for speech stimuli; normal ABR to 100 dB p.e. SPL clicks) (fig. 5) and a retrocochlear pattern (threshold for the acoustic stapedius reflex not attained at maximum audiometer output; absence of recruitment; speech discrimination score lower than 70% and/or marked

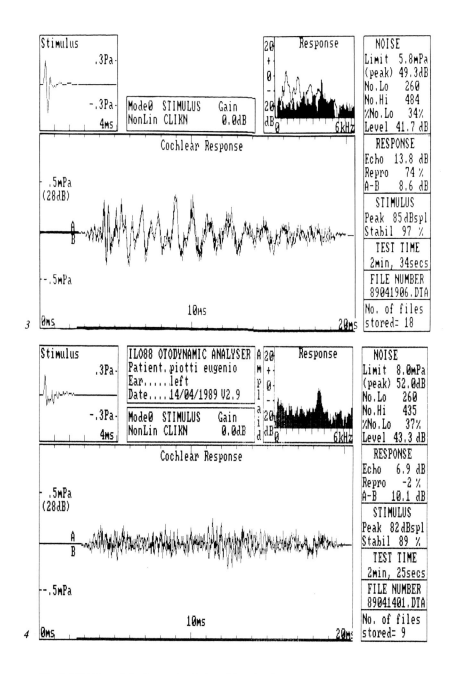

Fig. 3. EOAE to clicks in a noise-induced hearing loss case. Thresholds at 1, 2, 3, 4 kHz were 10, 10, 40, 55 dB HL. Method of Bray and Kemp [16].

Fig. 4. Absence of EOAE in a noise-induced hearing loss case. Thresholds at 1, 2, 3, 4 kHz were 15, 35, 45, 60 dB HL. Method of Bray and Kemp [16].

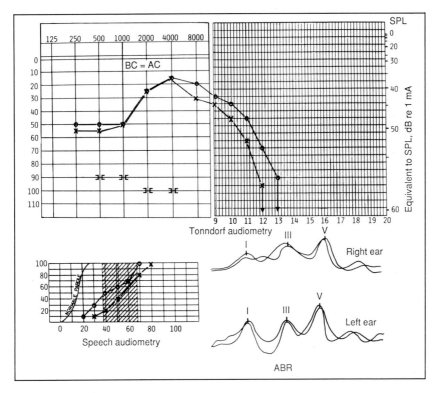

Fig. 5. Female (V.M., 58 years old) with the cochlear pattern of familial low-frequency hearing loss. EOAE were absent.

roll-over in the performance-intensity function for speech stimuli; abnormal ABR) (fig. 6).

EOAE were recorded in 3 out of 4 ears with the retrocochlear pattern only; they were absent in the ears with the cochlear pattern of 'familial low-frequency hearing loss'.

Conclusions

Detection of EOAEs in sensorineural hearing losses of the adult does not seem to add, per se, essential cues to the overall diagnostic picture [19].

It is more likely that in cases with subclinical changes of the cochlea, when pure-tone sensitivity is still within normal limits, the recording of

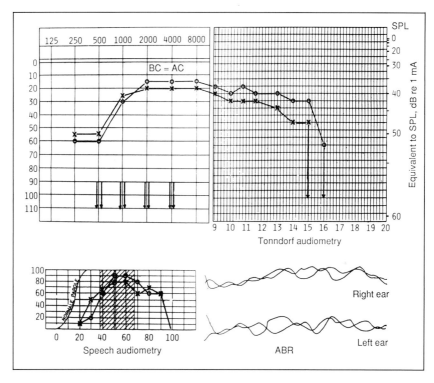

Fig. 6. Male (S.M., 23 years old) with the retrocochlear pattern of familial low-frequency hearing loss. EOAE were recorded from the right ear only.

EOAEs, and the analysis of their pattern and main parameters (latency, group latency, input/output function) can give more important information on the functional state of the end organ.

References

1   Kemp DT: Stimulated acoustic emissions from within the human auditory system. J Acoust Soc Am 1978;64:1386–1391.
2   Rutten WLC: Evoked acoustic emissions from within normal and abnormal human ears. Comparison with audiometric and electrocochleographic findings. Hear Res 1980;2:263–271.
3   Ruggero MA, Rich NC, Freyman DJ: Spontaneous and impulsively evoked oto-acoustic emissions: Indicators of cochlear pathology? Hear Res 1983;10:283–300.

4   Bonfils P, Uziel A, Pujol R: Evoked otoacoustic emissions: A fundamental and clinical survey. ORL 1988;50:212–218.
5   Johnsen NJ, Bagi P, Elberling C: Evoked acoustic emissions from the human ear. III. Findings in neonates. Scand Audiol 1983;12:17–24.
6   Bonfils P, Uziel A, Narcy P: Apport des émissions acoustiques cochléaires en audio-logie pédiatrique. Ann Otolaryngol 1988;105:109.
7   Tanaka Y: Stimulated otoacoustic emissions from ears with sensorineural hearing loss. Proc. Jpn Acad 1988;64B(7):201–204.
8   Tanaka Y, Inoue T, Tokita N: Continuity of otoacoustic emissions in ears with dip type hearing loss. Proc Jpn Acad 1988;64B (10):311–314.
9   Anderson SD, Kemp DT: The evoked cochlear mechanical response in laboratory primates. A preliminary report. Arch Otolaryngol 1979;224:47–54.
10 Johnsen NJ, Elberling C: Evoked acoustic emissions from the human ear. II. Nor-mative data in young adults and influence of posture. Scand Audiol 1982;11:69–77.
11 Grandori F, Ottaviani A, Ottaviani F, Tamplenizza P: Nonlinear derived evoked otoacoustic emissions in subjects with noise-induced hearing loss; in Quaranta A (ed): Clinical Audiology '87'. Bari, Laterza Edizioni Universitarie, 1988, pp 66–73.
12 Kemp D.T.: Cochlear echoes: Implications for noise-induced hearing loss; in Hen-derson D. Salvi RJ, Hamernick RA.: New Perspectives in Noise-Induced Hearing Loss. New York, Raven Press, 1982, pp 189–207.
13 Konigsmark BW: Mengel M, Berlin CI: Familial low-frequency hearing loss. Laryn-goscope 1971;81:759–771.
14 Grandori F: Evoked oto-acoustic emissions. Stimulus-response relationships. Rev Laryngol Otol Rhinol (Bordeaux) 1983; 104: 153–155.
15 Grandori F: Nonlinear phenomena in click and tone-burst evoked otoacoustic emis-sions from human ears. Audiology 1985;24:71–80.
16 Bray P, Kemp DT: An advanced cochlear echo technique suitable for infant screen-ing. Br J Audiol 1987;21:191–204.
17 Antonelli A, Grandori F: Long-term stability, influence of the head position and modelling considerations for evoked otoacoustic emissions. Scand Audiol 1986; 25(suppl):97–108.
18 Probst R, Coats AC, Martin GK, Lonsbury-Martin BL: Spontaneous, click- and tone-burst evoked otoacoustic emissions from normal ears. Hear Res 1986;21:261–275.
19 Kemp DT, Bray P, Alexander L, Brown AM: Acoustic emission cochleography. Practical aspects. Scand Audiol 1986;25(suppl):71–95.

V. Garrubba, Otorhinolaryngologic Clinic of the University, Spedali Civili,
I–25060 Brescia (Italy)

Grandori F, Cianfrone G, Kemp DT (eds): Cochlear Mechanisms and
Otoacoustic Emissions. Adv Audiol. Basel, Karger, 1990, vol 7, pp 156–163

Active Mechanisms and Cochlear Efferents

J.-L. Puel, G. Rebillard, R. Pujol[1]

Laboratoire de Neurobiologie de l'Audition, INSERM U-254,
Université de Montpellier II, Hôpital St. Charles, Montpellier, France

The otoacoustic emissions (OAEs) have a cochlear origin and reflect
an active process [1–3] which is responsible for the exquisite sensitive and
discriminative properties of the auditory receptor [4, 5]. A current hypothesis is that the fast motility of the isolated outer hair cells (OHCs) induced
by electrical stimulation is related to these active processes [6, 7], whereas
the chemically induced slow motility of the hair cells [8, 9] is related to the
modulation of the active process, modulation which is probably driven in
vivo by the efferent fibers. The only efferents connected to the OHCs are
the medial olivocochlear (MOC) fibers coming from the medial nuclei of
the superior olivary complex [for a review, see 10]. Consequently, any
variation of the OAEs observed after an efferent manipulation can be
exclusively attributed to the MOC efferents. Indeed, evidence that efferents alter OAEs exists. One type of OAEs, the acoustic distortion products
(DPs), is affected by an electrical stimulation of the olivocochlear bundle
(OCB) [11, 12]. Another type, the evoked otoacoustic emissions (EOAEs)
is significantly reduced in human subjects during a visual selective attentional task [13, 14]. Recently, Kemp and Souter [15] also showed that
OAEs were dynamically modulated by mild transient efferent stimulation.

[1] The authors are indebted to S. Ladrech and R. Chabert for their efficient technical
assistance. They would also like to thank E. Mayat for correcting the English, P. Sibleyras
for the photographical work and A. Bara for her careful editorial assistance.

On the other hand, it has been shown that the presentation of contralateral sound can influence the responses of the ipsilateral ear. Buno [16] and Murata et al. [17] reported that a moderate-level contralateral sound can suppress the responses of auditory nerve fibers to ipsilateral tones. They suggested that these effects were mediated by the efferents. In the same way, sound stimulation of contralateral ear attenuates the effect of acoustic trauma in the ipsilateral ear, apparently through efferent fibers [18, 19].

Therefore, we decided to test the hypothesis that a contralateral sound stimulation can activate the MOC efferents and modify OAEs. We chose to record the $2f_1-f_2$ DP, which is the most readily recordable OAE in guinea pigs.

Materials and Methods

Experiments were performed on 20 pigmented guinea pigs (200–400 g) of either sex. The animals were anesthetized with Urethane (1.4 mg/kg, i.p.), paralysed with Flaxedil and artificially ventilated. The rectal temperature was maintained at $38.5 \pm 1\,°C$. Supplementary doses of Urethane were administered as needed. Heart rate was monitored via ECG electrodes. The lowest part of the pinna was excised to expose the external auditory canal. The right cochlea of each animal was exposed ventrolaterally and the stapedius and tensor tympani muscles were severed. After a posterior craniotomy, the cerebellum was removed by aspiration to expose the floor of the fourth ventricle. The measurements reported in this paper were done before and after medially sectioning the brainstem at the floor of the fourth ventricle, in a rostrocaudal direction.

The stimulus frequencies f_1 and f_2 (primary tones) were generated by a custom-made frequency synthesizer. The primary tones were delivered to the sound delivery-receiving device. It consisted of a Knowles microphone (1751) and of two Knowles emitters (1850) (which were connected to a plastic coupler). The probe was fitted within the right external ear canal to form a closed acoustic system. In the present study, the $2f_1-f_2$ DP was investigated at 5 kHz and the f_2/f_1 ratio was equal to 1.17. The magnitude of the $2f_1-f_2$ DPs was measured at an intensity of the primaries of 60 dB SPL. At the end of each experiment, the guinea pigs were sacrificed with an overdose of Nembutal. In all cases, the $2f_1-f_2$ DP value fell to the background noise level within 2 min after the cardiac arrest, attesting to the physiological origin of the DP.

The contralateral stimulation was generated by a Brüel & Kjaer white noise generator, passed through an attenuator and applied in a closed field to the left auditory canal via an ear phone (JBL 075) and a catheter. During the experiment the white noise was delivered to the left ear at 100, 60 and 20 dB SPL. No 'cross-talk' between the left and the right ear was recorded at these intensities of stimulation.

Calibration of the acoustic system was carried out in an artificial ear by using a calibrated ½-inch Brüel & Kjaer microphone (4134) and a Brüel & Kjaer measure amplifier (2606). Both emitters, and microphones were calibrated for all frequencies used. The

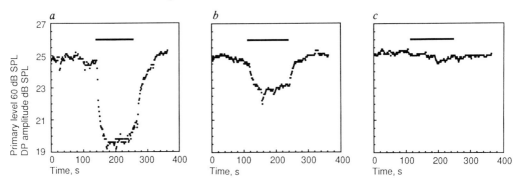

Fig. 1. Example of DP reduction during contralateral white noise stimulation. The reduction of the DP $2f_1-f_2$ recorded for a primary intensity of 60 dB SPL varies with the intensity of the contralateral stimulation (horizontal bar). The reduction is greater with 100 dB SPL (*a*) than with 60 dB SPL (*b*), no reduction is seen with 20 dB SPL (*c*).

artifactual distortion of the probe ($2f_1-f_2$), tested in a passive cavity of the same volume as guinea pig ears, was 75–80 dB below the primaries.

The ear-canal sound pressure was directly analyzed by a Brüel & Kjaer 2033 spectrum analyzer which provided a real-time Fourier analysis. For $2f_1-f_2$ DP measurements, analysis bandwidth was typically 7.5 Hz with a Hanning window and a frequency span of 2 kHz. Each actual measurement was the exponential average of the last eight fast Fourier transforms (FFT). This FFT average was collected every 1.5 s. With this equipment and depending on experiments, our noise floor was measured between –6 and –12 dB SPL ranging from 1 to 20 kHz.

The DP was measured as a function of time during 400 s. During this recording the contralateral white noise was applied for approximately 100 s. This procedure was repeated for 3 intensities of white noise. The floor of the fourth ventricle was then sectioned and the recording procedure repeated. The results are expressed in decibels as the mean ± SEM.

Results

The mean magnitude of the DPs measured in dB SPL at 5 kHz was 26.9 ± 1.31 for a primary level of 60 dB SPL. In the tested ears, the stability of the DPs, as a function of time was very good (less than ± 0.3 dB in 4 h).

In all animals, the presentation of contralateral white noise induced a clear reduction of the DP magnitude which had been evoked by primaries delivered at an intensity of 60 dB SPL. The amount of reduction was a

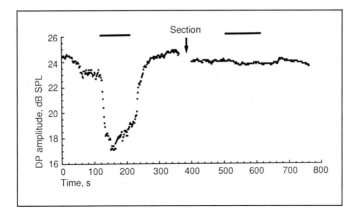

Fig. 2. Effect of a section of the floor of the fourth ventricle. The effect of the same contralateral white noise stimulation (horizontal bars) is shown before and after a total section of the floor of the fourth ventricle (arrow). This section completely eliminates the white noise reducing effect.

function of the intensity of the contralateral white noise. The mean reduction expressed in dB SPL was 3.32 ± 0.61 for an intensity of contralateral white noise of 100 dB SPL; this value decreases to 1.32 ± 0.23 for 60 dB SPL contralateral white noise. No variation appeared with a contralateral sound of 20 dB SPL (fig. 1).

A section of the midbrain at the floor of the fourth ventricle had no effect on DP magnitude. However, this section completely blocked the suppressive effect of the contralateral white noise, even when applied at 100 dB SPL (fig. 2).

Discussion

The present study shows that contralateral sounds can reduce the magnitude of the $2f_1-f_2$ DP recorded in the ipsilateral ear and that this suppressive effect disappears after sectioning of all the fibers crossing the brainstem midline at the floor of the fourth ventricle.

Evidence that a contralateral acoustic stimulation can affect ipsilateral responses exists in the literature. For example, it has been shown that the presentation of moderate-level contralateral sound can suppress the responses of auditory-nerve fibers to ipsilateral tones at characteristic frequencies [16, 17]. Such a suppression completely disappeared following the interruption of the efferents at the vestibulocochlear anastomosis [20]. In addition, sound stimulation of the contralateral ear attenuates the effect of acoustic trauma on the ipsilateral ear, apparently through efferent fibers [18, 19]. However, some of these studies do not provide definite proof that the middle-ear muscles do not contribute to the observed binaural interaction, and most of them do not clearly show whether this phenomenon can be attributed to the MOC efferents.

In the present study, all animals tested were paralysed with Flaxedyl and the stapedius and tensor tympani muscles of the tested ear were cut. In addition, the contralateral acoustic stimuli were restricted to levels such that no 'cross-talk' between ears could be identified in the auditory canal, i.e., no variation of the noise floor level (-6 to -12 dB SPL from 1 to 20 kHz). Thus, the possiblity that middle-ear acoustic reflex and/or interaural acoustic cross-talk would be responsible for the effects described in this report can be ruled out.

The cochlea is innervated by two efferent components: (1) the lateral olivocochlear (LOC) efferents which originate in the area of the lateral superior olive and synapse with the dendrites of the auditory nerve below the inner hair cells (IHCs), and (2) the MOC efferents which originate in the medial nuclei of the superior olivary complex and synapse on the OHCs [for a review, see ref. 10]. The main interest of the present results is to specify which efferent system (LOC or MOC) is involved in the suppressive effect by the contralateral sound stimulation. Actually, the DPs, which reflect the nonlinearities of the cochlea, are considered to be the result of an active biomechanism expressed through the OHCs [see, among others, refs. 4, 5, 21–23]. These OHCs receive only MOC efferents. Consequently, any variation of the DPs occurring after an electrical stimulation of the OCB [11, 12] or, during contralateral sound stimulation, as we reported herein, can be exclusively attributed to an activation of the MOC efferents. In addition, sectioning all the fibers crossing at the midline of the floor of the fourth ventricle completely eliminates the suppressive effect of the contralateral sound stimulation. Unfortunately, this type of section, which is not selective and interrupts both afferent and efferent pathways, cannot tell us which part of the MOC efferents (crossed or uncrossed) is activated during contralateral sound stimulation. Nevertheless, the present study clearly shows that a physiological stimulus, such as a contralateral sound stimulation, can activate the MOC efferents and alter the cochlear micromechanics of the ipsilateral cochlea. This is probably the way by which efferents attenuate the intracellular IHC responses [24] and elevate the threshold of auditory nerve fibers [25].

In conclusion, it has been previously suggested that the efferents could be involved in the ipsilateral masking reinforcement phenomenon [26–28]. The present study clearly demonstrates that a contralateral masker can also activate the MOC efferents to change the micromechanics of the ipsilateral ear. This result is consistent with those of Cody and Johnstone [18] and Rajan and Johnstone [19] who showed that a contralateral ear acoustic stimulation can reduce the amount of desensitization produced by an intense sound. Since this effect was blocked by an intramuscular injection of strychnine, these authors proposed that the desensitization could be the result of an acoustically evoked contralateral efferent activity. Thus, our results support this protective effect of a contralateral sound stimulation against acoustic trauma, an effect which is mediated through the MOC efferents. Moreover, the MOC efferents can also be activated by the upper central nervous system during an attentional task [13, 14]. All together, these findings support the idea that an evaluation of the functional state of the MOC efferents may be of clinical interest. Since OAE DPs as well EOAEs can be easily recorded in humans [see, among others, refs. 3, 29–31], the contralateral sound suppression of OAEs would be useful to investigate the activity of MOC efferents in humans.

References

1 Kemp DT: Stimulated acoustic emissions from the human auditory system. J Acoust Soc Am 1978;64:1386–1391.
2 Kemp DT: Otoacoustic emissions, travelling waves and cochlear mechanisms. Hear Res 1986;22:95–104.
3 Wilson JP: Evidence for a cochlear origin for acoustic re-emissions, threshold fine-structure and tonal tinnitus. Hear Res 1980;2:233–252.
4 Davis H: An active process in cochlear mechanics. Hear Res 1983;9:79–90.
5 Neely ST, Kim DO: An active cochlear-model showing sharp tuning and high sensitivity. Hear Res 1983;9:123–130.
6 Brownell WE, Bader CR, Bertrand D, de Ribaupierre Y: Evoked mechanical responses of isolated cochlear outer hair cells. Science 1985;227:194–196.
7 Ashmore JF: A fast motile response in guinea-pig outer hair cells: The cellular basis of the cochlear amplifier. J Physiol 1987;388:323–347.
8 Zenner HP: Motile responses in outer hair cells. Hear Res 1986;22:83–90.
9 Flock Å, Flock B, Ulfendahl M: Mechanisms of movement in outer hair cells and a possible structural basis. Arch Oto-Rhino-Laryngol 1986;243:84–91.

Grandori F, Cianfrone G, Kemp DT (eds): Cochlear Mechanisms and
Otoacoustic Emissions. Adv Audiol. Basel, Karger, 1990, vol 7, pp 164–170

Influence of a Contralateral Auditory Stimulation on Evoked Otoacoustic Emissions

L. Collet[a, b], *E. Veuillet*[a], *R. Duclaux*[b], *A. Morgon*[a]

[a] Laboratoire d'Exploration Fonctionnelle Neurosensorielle, Pavillon U,
Hôpital Edouard-Herriot; Lyon; [b] Laboratoire de Physiologie Sensorielle,
Faculté de Médecine Lyon-Sud, France

In an initial work, Buno [1] has demonstrated that contralateral auditory stimulation could elicit changes in spontaneous auditory nerve activity. These results arouse two kinds of question: (1) a physiological one: Is there a mutual cochlear interaction? and (2) a audiological one: Do we have a model for functional exploration of the efferent system? The aim of this paper is to present new data about these questions using otoacoustic emissions (OAEs).

Is There a Mutual Interaction between the Two Organs of Corti?

Buno [1] has recorded spontaneous auditory nerve activity in cats with and without a low-intensity contralateral forward stimulation. A decrease in responses under contralateral pure tone stimulation and an increase under contralateral white noise are observed. These experiments have been partially replicated by Warren and Liberman [2, 3] who found a suppression of the responses of auditory nerve afferents whatever the contralateral sound tones and broadband noise. Furthermore, the suppressive effect of contralateral stimulation disappeared after section of the olivo-cochlear bundle. Other studies support this cochlear interaction: Rajan and Johnstone [4] showed a reduction in the cochlear action potential temporary threshold shift (TTS) in the presence of contralateral acoustic stimulation. More recently (1988), the same authors [5] have demonstrated on TTS an analogy between the effect of binaural acoustic stimulation and electrical stimulation of auditory efferent fibers.

In humans, a fall in compound action potential amplitude (N_1) with contralateral pure sound stimulation has been shown by Folsom and Osley [6].

A histological study has confirmed the possibility of an interaction between the two cochleas. Dodson et al. [7] have electrically stimulated (extracochlear stimulation) one cochlea in guinea pigs. They have shown structural effects in both ears especially at the cochlear efferent endings in terms of a bilateral increase in the number of synaptic vesicles.

For all these authors, the contralateral effect is mediated by the efferent system. In fact, two distinct efferent systems must be considered [8]: the medial efferent system – essentially contralateral – innervates directly the outer hair cells while the lateral efferent system – essentially ipsilateral – does not innervate the hair cells but the type I cochlear afferent neurons near the inner hair cells. So, the latter system acts directly onto the auditory afferents. The suppression of the response of auditory nerve afferents under contralateral auditory stimulation may be the consequence either of a direct effect (via the lateral efferent fibers) or of an indirect effect (via the medial efferent fibers) or of both. So all the studies mentioned above do not allow one to distinguish between the respective roles of the two efferent systems. Low-level acoustic signals of cochlear origin can be recorded in the ear canal: the otoacoustic emissions [9]. they are generated by outer hair cell motility. The interest in the use of OAEs is this cochlear specificity. Outer hair cells involved in the genesis of OAEs and innervated by the medial efferent system, their modifications under contralateral auditory stimulations are a proof of a mutual interaction.

Alteration of Otoacoustic Emissions by Contralateral Auditory Stimulation

To our knowledge, only four studies deal with this subject. In the earliest papers [10, 11], distortion products have been used. The remaining two papers deal with spontaneous OAEs [12] and evoked OAEs [14], respectively.

Brown [13] described an alteration at the level of acoustic distortions during long periods of continuous stimulation. A contralateral stimula-

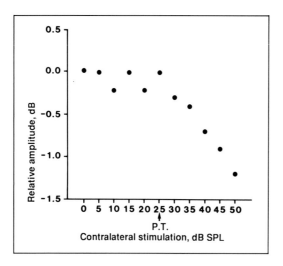

Fig. 1. Relative amplitude of OAE according to contralateral stimulation in one typical subject (nonobstructed ear). PT = Perception threshold.

tion during the period of rest decreases the recovery of initial distortion level.

Puel et al. [11] have shown a significant decrease of otoacoustic distortion products under white noise contralateral stimulation and the disappearance of this effect after section of the floor of the fourth ventricle.

In humans, Mott et al. [12] reported that spontaneous OAEs shift upward in frequency under contralateral tonal stimulation without significant amplitude changes. Finally, Collet et al. [14] have demonstrated a reduction of OAE amplitude under simultaneous low-intensity contralateral white noise. Figure 1 shows the results from a representative subject showing a decrease of OAE amplitude which is greater when contralateral stimulus intensity increases. 10 dB above the threshold are sufficient to elicit the contralateral effect.

To discuss these results in terms of a mutual cochlear interaction, a technical artifact, a cross-talk attenuation and a middle-ear role must be eliminated. To rule out any technical artifact, the study has been replicated with and without obstruction of the contralateral ear (ear plugs). Figure 2 shows the results obtained in one subject after obstruction: the effect of the contralateral auditory stimulation disappears below 60 dB SPL. An effect is found only when the level of the contralateral sound is greater than 10

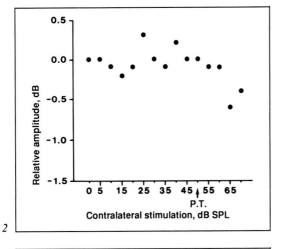

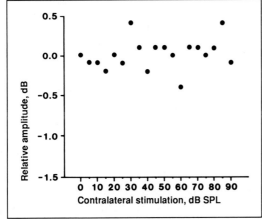

Fig. 2. Relative amplitude of OAE according to contralateral stimulation in one typical subject with obstruction of outer auditory canal. PT = Perception threshold.

Fig. 3. Relative amplitude OAE according to contralateral stimulation in a unilaterally cophotic subject.

dB SL. This experiment allows to eliminate a technical artifact due to air conduction.

To rule out the cross-talk attenuation, OAE of unilateral cophotic subjects have been recorded with and without contralateral stimulation of the cophotic ear. No contralateral auditory stimulation effect on OAE intensity was found (fig. 3). This result eliminates an interaural bone transmis-

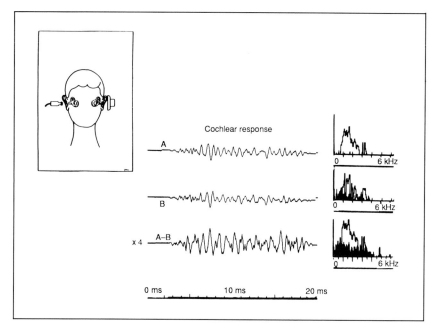

Fig. 4. OAE and spectral analysis of a typical subject in the absence (A) or the presence (B) of a 50-dB contralateral white noise stimulus.

sion effect. The implication of middle ear is more difficult to eliminate. However, it must be noted that the effect of contralateral white noise appears at 10 dB SL, an intensity too low to elicit the acoustic reflex. Moreover, the digital subtraction of the OAE curves (fig. 4) with and without contralateral stimulation shows that almost all the frequencies of the OAE are decreased under contralateral stimulation and not only low frequencies. Kemp demonstrated that an alteration of middle ear pressure acts essentially on low-frequency components of OAE and less on high frequencies. These two last results provide indirect evidence against a middle-ear role. A direct proof is the influence of contralateral auditory stimulation on evoked OAE of patients without acoustic reflex. Figure 5 shows the fall-off in OAE intensity in a case of Bell's palsy (without white noise acoustic reflex). This effect being the same as in control subjects, the role of facial nerve is eliminated. Therefore, these results provide some evidence that contralateral auditory stimulation can modify evoked OAE. This

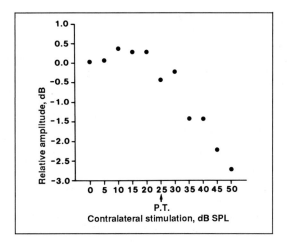

Fig. 5. Relative amplitude of OAE according to the contralateral stimulation in a case of Bell's palsy (without white-noise acoustic reflex).

effect cannot be explained in terms of technical artefact, cross-talk, and middle-ear effect. It may be the consequence of an active cochlear micro-mechanism via the medial efferent system.

From the evidence reviewed here, it is clear that recording of OAE with and without contralateral white noise stimulation can help to functionally explore the medial olivocochlear efferent system. This technique is easy, objective, and noninvasive, but its main limit is the absence of OAE in pathological ears [15].

To conclude, the mutual interaction between the two organs of Corti has been proved and there is a model for functional exploration of the medial efferent system for physiological and clinical purposes.

References

1   Buno W: Auditory nerve activity influenced by contralateral ear sound stimulation. Exp Neurol 1978;59:62–74.
2   Warren EH III, Liberman MC: Effects of contralateral sound on auditory-nerve responses. I. Contributions of cochlear efferents. Hear Res 1989;37:89–104.
3   Warren EH III, Liberman MC: Effects of contralateral sound on auditory-nerve responses. II. Dependence on stimulus variables. Hear Res 1989;37:105–122.

4   Rajan R, Johnstone BM: Crossed cochlear influences on monaural temporary
     threshold shifts. Hear Res 1983;9:279–294.
5   Rajan R, Johnstone BM: Binaural acoustic stimulation exercises protective effects at
     the cochlea that mimic the effects of electrical stimulation of an auditory efferent
     pathway. Brain Res 1988;459:241–255.
6   Folsom RL, Owsley RM: N_1 action potentials in humans. Influence of simultaneous
     contralateral stimulation. Acta Otolaryngol (Stockh) 1987;103:262–265.
7   Dodson HC, Walliker JR, Bannister LH, et al: Structural effects of short term and
     chronic extracochlear electrical stimulation on the guinea pig spiral organ. Hear Res
     1987;31:65–78.
8   Warr WB, Guinan JJ Jr: Efferent innervation of the organ of Corti: Two separate
     systems. Brain Res 1978;173:152–155.
9   Kemp DT: Stimulated acoustic emissions from within the human auditory system. J
     Acoust Soc Am 1978;64:1386–1391.
10   Brown MC, Nuttal AL, Masta RI: Intracellular recordings from cochlear inner hair
     cells: Effects of stimulation of the crossed olivo-cochlear efferents. Science 1983;
     222:69–72.
11   Puel JL, Rebillard G, Pujol R: Active mechanisms and cochlear efferents. Proc 2nd
     Int Symp Cochlear Mechanics and Otoacoustic Emissions, Rome 1989. Adv Audiol.
     Basel, Karger, 1990, vol 7, pp 156–163.
12   Mott JB, Norton SJ, Neely ST et al: Changes in spontaneous otoacoustic emissions
     produced by acoustic stimulation of the contralateral ear. Hear Res 1989;38:229–
     242.
13   Brown AM: Continuous low level sound alters cochlear mechanics: An efferent
     effect? Hear Res 1988;34:27–38.
14   Collet L, Kemp DT, Veuillet E, et al: Effect of contralateral auditory stimuli on
     active cochlear micro-mechanical properties in human subjects. Hear Res. In
     press.
15   Collet L, Gartner M, Moulin A, et al: Evoked otoacoustic emissions and sensori-
     neural hearing-loss. Arch Otolaryngol Head Neck Surg. 1989;115:1060–1062.

L. Collet, MD, Laboratoire d'Exploration Fonctionnelle Neurosensorielle,
Hôpital Edouard-Herriot, Pavillon U, place d'Arsonval, F–69374 Lyon (France)

Grandori F, Cianfrone G, Kemp DT (eds): Cochlear Mechanisms and
Otoacoustic Emissions. Adv Audiol. Basel, Karger, 1990, vol 7, pp 171–179

Evoked Otoacoustic Emissions and Cochlear Psychoacoustic Tests: Effects of Atropine[1]

Antonio Quaranta, Ignazio Salonna

Center of Audiology and Otology, ENT Department, University of Bari, Italy

An efferent innervation of the cochlea has been known to exist since
Rasmussen [1] demonstrated a homo- and contralateral efferent olivocochlear bundle in cats. Small unmyelinated fibers of this bundle lead to the
area of the inner ear cells, making numerous synaptic contacts 'en passant'
with the afferent inner radial fibers of the inner hair cell system [2], while
other fibers provide a very abundant efferent innervation of the outer hair
cells [3, 4], having an enormous contact area with their body [5, 6].

Experimental research in animals has shown that stimulation of the
crossed olivococochlear bundle reduces the action potential of the eighth
nerve [7, 8] and the receptor potential of both inner (IHCs) [9] and outer
[10] hair cells (OHCs). Mountain [11] and Siegel and Kim [12] measured
distortion products in the ear canal produced by two tones and found that
these distortion products were changed by stimulation of olivocochlear
efferents. Since this efferent effect on distortion products could be blocked
by perfusing the cochlea with curare [12], Guinan [10] speculated that
stimulation of the olivocochlear efferent fibers produces a mechanical
change in the cochlea. On the other hand, the anatomic section of the
crossed olivocochlear bundle leads to an increase in cochlear frequency
selectivity without impairments of tonal and masked thresholds [13–15].

It has been known for many years that the neurotransmitter for the
efferent synaptic system is acetylcholine (ACh); in fact, it was shown that

[1] This work was supported in part by grant from CNR – Consiglio Nazionale delle
Ricerche (Project on preventive and rehabilitative medicine) and from Ministero della
Pubblica Istruzione.

enhancement of the amplitude of the cochlear microphonic-auditory nerve and reduction of the action potential following electrical stimulation of the crossed olivocochlear bundle [16] are blocked by endocochlear injection of nicotinic ACh antagonists such as hexamethonium and succinylcholine [17], and muscarinic antagonists, such as atropine [18].

In the last few years, polymerized actin has been identified as the basic cytoskeleton of the stereocilia and accounts for their stiffness [19, 20]. More recently, a number of investigators have observed motile activity of dissociated OHCs but not of dissociated IHCs [21–23]. Although the exact cellular mechanisms for OHC motility is not fully understood, a number of important cytoskeletal and contractile proteins have been observed along the inner periphery of the OHC membrane [myosin: ref. 24, 25; actin: ref. 26] and in the cuticular plate into which the rootlets of stereocilia insert [nonpolymerized actin: ref. 22, α-actinin, fimbrin, and myosin: ref. 27, 28]. This complex cytoskeletal organization would strongly argue in favor of both the possibility that the OHCs can vibrate through the molecular coupling of the actomyosin complex [24] and of an active role of the cuticular plate in regulating the stiffness of the stereociliary bundles [26].

Some experiments have suggested two different ways in which efferents on hair cells might produce a mechanical change of the OHCs.

First, Brownell et al. [21], using an in vitro preparation, showed that the application of ACh shortens the OHCs. Second, Ashmore [29] reported that electrical stimulation, which excites efferent fibers to hair cells in the frog's sacculus, produces a change in the position of the hair cell stereocilia.

The aim of the present research was to evaluate, in normal-hearing ears, the effects of atropine on the performance of advanced tests of cochlear function in order to analyze the consequences of the functional inhibition of the efferent system on cochlear efficiency.

Material and Methods

Subjects

We studied 10 ears of 10 young subjects (6 females), ranging in age from 18 to 35 years; they were volunteers and selected according to the following criteria: normal cardiac function; no history of otological, neurological and internal diseases, noise exposure, head trauma, use of ototoxic drugs or any other product that could interfere with normal function of the central nervous system; hearing within 15 dB HL for all frequencies up to 8,000 Hz (ISO, 1975); type A tympanogram and acoustic reflex threshold ranging between

75 and 90 dB SPL; normal cochlear performance evaluated by remote masking (RM), critical ratio (CR) and evoked otoacoustic emissions (EOAEs). The test results were recorded before and after intravenous infusion of 1 mg of atropine.

Equipment

The apparatus used for our study consisted of: (a) two dual-channel pure-tone audiometers (Mercury M-132); (b) an EM-40 Hearing Science Laboratory, built by EAC-Mercury; (c) a clinical averaging system Amplaid MK7, interfaced with an electroacoustic bridge equipped with an x-y plotter for tympanometry and acoustic reflex measurements (Amplaid 702) and with small acoustic probe, manufactured by Amplaid incorporating a miniature earphone (Knowles BT-2606) and a condenser microphone (Knowles BT-1751).

Procedure

Remote Masking. It has been proposed as a test of cochlear partition rigidity [30], and consists of a threshold increase for low-frequency tones when the ear is exposed to a high-frequency noise band delivered at an overall noise level of 80–100 dB SPL. RM was recorded ipsilaterally for pulsed tones (duration 250 ms, rise-fall time 25 ms, duty cycle 50%) at 250 Hz using a continuous narrow-band noise masker centered at 3,000 Hz with a bandwidth of 305 Hz, and delivered at an overall level of 98 dB SPL.

RM values ranged in normal ears between 11 and 35 dB; values below 11 dB were considered pathological.

Critical Ratio. This was calculated according to the formula [31]:

CR (dB) = masked threshold level – spectrum level of masking noise.

After Fletcher [32], several authors utilize CR as an indirect evaluation of the critical bandwidth and therefore of the frequency selectivity of the auditory system.

Physiological and psychological evidence [33] suggests that this most important function is largely accomplished at the level of the cochlea, depending on the filtering process both of the basilar membrane (primary filter) and of the organ of Corti, particularly of the OHCs (hypothetical secondary filter).

CR was calculated ipsilaterally for 1,000-Hz pulsed tones (duration 500 ms, rise-fall time 25 ms, duty cycle of 50%). The masking stimulus was a continuous wide-band noise ranging from 90 to 20,000 Hz, delivered at an overall level of 83 dB SPL with a spectrum level of 40 dB. CR values range from 16 to 21 dB and any CRs above 21 dB were considered abnormally increased.

Evoked Otoacoustic Emissions. As is well known [34], these cochlear echo responses, evoked by transient acoustic stimulation, are a property of the healthy functioning cochlea [35, 36]. The mechanism of EOAE generation is understood in general terms only.

The first hypothesis by Kemp [34, 37] and Kim [38] suggested that EOAEs arise from reflections from a mechanical impedance discontinuity in the basilar membrane, but recent direct observations of cochlear mechanisms lead to believe that all parts of the cochlea, included, presumably, OHCs are involved in the generation of EOAEs [35, 39, 40].

EOAEs have been recorded in our subjects, sending 2,048 clicks at a rate of 21 pps, with a time window of 30 ms and filtered with a band-pass (200–5,000 Hz), with a stim-

therefore that the integrity of the cholinergic efferent system of the cochlear nucleus was necessary for normal frequency resolution.

The increase of CR values observed in our cases is similar to Pickles and Comis' [41] and Pickles' [42] findings, and indicates that also in man, the cochlear filtering process depends on the functional state of the efferent auditory pathways. The influence of the cholinergic system on cochlear efficiency emerges also considering the post-atropine decrease of RM and of EOAE detection thresholds.

The details of the peripheral terminals of the efferent auditory pathways between the olivocochlear bundle and the OHCs and the new concepts of auditory sensory transduction are well known [3, 4]: they suggest an active role of the OHCs because of the contractile activity of their cellular body and stereocilia [5, 26]. In addition, Kim [43] suggested that the OHCs can influence the basilar membrane motion.

Based on these physiological data and keeping in mind Pickles and Comis' [41] and Pickles' [42] conclusions, our results seem to confirm the cochlear origin of RM, CR and EOAEs, and lead to think that OHCs play an important role in producing these three phenomena. Such a hypothesis has already been suggested for CR [33] and EOAEs [37, 39] but is new for RM, which has been attributed, until now, only to a mechanical nonlinear distortion of the cochlear partition [44–46].

It is difficult to explain the differences in behavior observed in our subjects between CR and EOAEs after the injection of atropine. In fact the CR values were increased with a consequent reduction of cochlear frequency selectivity, whereas EOAEs presented a decrease in the detection threshold, thus showing a facilitation on cochlear echo response. We could assume that pharmacologic block of the centrifugal pathways causes an increase of mechanical impedance of OHCs which, on the one hand, accept less energy to be transduced into a neural stimulus, making masked thresholds and consequently CRs increase, and, on the other hand, rejects more energy thus decreasing the EOAE detection threshold.

Conclusions

The findings in tests of cochlear efficiency in normal-hearing subjects recorded before and after injection of atropine show that the pharmacologic block of centrifugal pathways causes an impairment of cochlear performance, hypothetically due to an increase of the mechanical impedance of OHCs.

References

1   Rasmussen GL: An efferent cochlear bundle. Anat Rec 1942;82:441.
2   Warr WB: The olivocochlear bundle: its origins and termination in the cat; in Fernandez C, Naunton NF (eds): Evoked electrical activity in the auditory nervous system. New York, Academic Press, pp 43–65, 1978.
3   Spoendlin H: Efferent innervation of the cochlea; in Bolis L, Kergues RD, Maddrell SHP (eds): Comparative physiology of the sensory system. Cambridge, Cambridge University Press, 1984.
4   Warr WB, Guinan JJ: Efferent innervation of the organ of Corti: Two separate systems. Brain Res 1979;173:152–155.
5   Flock Å: Hair cells, receptors with a motor capacity? in Klinke R, Hartman R (eds): Hearing – Physiological Bases and Psychophysics. Berlin, Springer, 1983, pp 1–6.
6   Takasaka T, Shinkawa H, Hashimoto S, Watanuki K, Kawamoto K: High-voltage electron microscopic study of the inner ear. Ann Otol Rhinol Laryngol 1983; 92(suppl 101):1–12.
7   Fex J: Efferent inhibition in the cochlea related to hair cell activity: Study of postsynaptic activity of the crossed olivocochlear fibres in the cat. J Acoust Soc Am 1967;41:666–675.
8   Galambos R: Suppression of auditory nerve activity by stimulation of efferent fibers to cochlea. J Neurophysiol 1956;19:424–437.
9   Brown MC, Nuttal AL, Masta RI: Intracellular recordings from cochlear inner hair cells: Effects of stimulation of the crossed olivocochlear efferents. Science 1983;222: 69–72.
10   Guinan JJ Jr: Effect of efferent neural activity on cochlear mechanics. Scand Audiol 1986;25(suppl):53–62.
11   Mountain DC: Changes in endolymphatic potential and crossed olivo-cochlear bundle stimulation alter cochlear mechanics. Science 1980;210:71–72.
12   Siegel JH, Kim DO: Efferent neural control of cochlear mechanics? Olivocochlear bundle stimulation effects cochlear biomechanical nonlinearity. Hear Res 1982;6: 171–182.
13   Capps MJ, Ades HW: Auditory frequency discrimination after transection of the olivocochlear bundle in squirrel monkeys. Exp Neurol 1968;21:147–158.
14   Galambos R: Studies of the auditory system with implanted electrodes; in Rasmussen GL, Windle WF, (eds): Neural Mechanisms of the Auditory and Vestibular System. Springfield, Thomas, 1960, pp 137–151.
15   Trahiotis C, Elliott DN: Behavioral investigation of some possible effects of sectioning the crossed olivo-cochlear bundle. J Acoust Soc Am 1970;47:592–596.
16   Klinke R, Galley N: Efferent innervation of vestibular and auditory receptors. Physiol Rev 1974;54:316–357.
17   Galley N, Klinke R, Oertel W, Pause MM, Storch WH: The effect of intracochlearly administered acetylcholine-blocking agents on the efferent synapses of the cochlea. Brain Res 1973;64:55–66.
18   Robertson D, Johnstone BM: Efferent transmitter substance in the mammalian cochlea: Single neuron support for acetylcholine. Hear Res 1978;1:31–36.
19   Flock Å: Contractile proteins in hair cells. Hear Res 1980;2:411–412.

20 Flock Å, Cheung HC: Actin filaments in sensory hairs of inner ear receptor cells. J Cell Biol 1977;75:339–343.

21 Brownell WE, Boder CR, Bertrand D, de Ribaupierre Y: Evoked mechanical responses of isolated cochlear outer hair cells. Science 1985;227:194–196.

22 Flock Å, Cheung HC, Flock B, Utter G: Three sets of actin filaments in sensory cells of the inner ear. Identification and functional orientation determined by gel electrophoresis, immunofluorescence and electron microscopy. J Neurocytol 1981;10:133–147.

23 Zenner HP, Zimmermann V, Schmitt V: Reversible contraction of isolated mammalian cochlear hair cells. Hear Res 1985;18:127–133.

24 Flock Å: Contractile and structural proteins in the auditory organ; in Dresher DG (ed): Auditory Biochemistry. Springfield, Thomas, 1985, pp 310–316.

25 Flock Å, Flock B, Ulfendahl M: Mechanisms of movement in outer hair cells and a possible structural basis. Arch Otorhinolaryngol 1986;243:83–90.

26 Lim DJ: Cochlear micromechanics in understanding otoacoustic emission. Scand Audiol 1986;25(suppl):17–25.

27 Drenckhahn D, Schafer T, Prinz M: Actin, myosin, and associated proteins in the vertebrate auditory and vestibular organs: Immunocytochemical and biochemical studies; in Drescher DG (ed): Auditory Biochemistry. Springfield, Thomas, 1985, pp 317–335.

28 Flock Å, Hoppe Y, Way X: Immunofluorescence localization of proteins in semithin 0.1–1 μm frozen sections of the ear: A report of improved techniques including gelatin encapsulation and cryoultramicrotomy. Arch Otorhinolaryngol 1981;233:55–66.

29 Ashmore SF: The stiffness of the sensory hair bundle of frog saccular hair cells. J Physiol 1984;350:20P.

30 Cervellera A, Quaranta A, Cassano P: Le remote masking: un test de conduction cochléaire. Audiology 1978;17:317–323.

31 Zwicker E: Über psychologische und methodische Grundlagen der Lautheit. Acustica 1958;8:237–258.

32 Fletcher H: Auditory pattern. Rev Mod Phys 1940;12:47–65.

33 Evans EF: Recent advances in Cochlear Physiology; in Gibb A, Amith MFW (eds): Otology. London, Butterworths, 1982, pp 104–131.

34 Kemp DT: Stimulated acoustic emissions from within the human auditory system. J Acoust Soc Am 1978;64:1386–1391.

35 Wit HP, Ritsma RJ: Stimulated acoustic emissions from the human ear. J Acoust Soc Am 1979;66:911–913.

36 Ruggero MA, Rich NC, Freyman R: Spontaneous and impulsively evoked otoacoustic emissions: Indicators of cochlear pathology? Hear Res 1983;10:282–300.

37 Kemp DT: Towards a model for the origin of cochlear echoes. Hear Res 1980;2:533–548.

38 Kim DO, Molnar CE, Matthews JW: Cochlear mechanics: Non-linear behavior in two-tone responses as reflected in cochlear nerve-fibre responses and in ear canal sound pressure. J Acoust Soc Am 1980;67:1704–1721.

39 Wilson JP: Otoacoustic emissions and hearing mechanism. Revue Laryngol 1984;105:179–191.

40 Wilson JP: Otoacoustic emissions and tinnitus. Scand Audiol 1986;25(suppl):109–120.
41 Pickles JO, Comis SD: Role of centrifugal pathways to cochlear nucleus in detection of signals in noise. J Neurophysiol 1973;36:1131–1137.
42 Pickles JO: Role of centrifugal pathways to cochlear nucleus in determination of critical bandwidth. J Neurophysiol 1976;39:394–400.
43 Kim DO:. Functional roles of the inner and outer-hair-cell subsystems in the cochlea and brainstem; in Berlin CI (ed): Hearing Science: Recent Advances. San Diego, College-Hill Press, 1984, pp 241–262.
44 Deatherage BH, Bilger RC, Eldredge DH: Remote masking in selected frequency regions. J Acoust Soc Am 1957;29:512–514.
45 Deatherage BH, Bilger RC, Eldredge DH: Physiological evidence of the masking of low frequencies by high. J Acoust Soc Am 1957;29:132–137.
46 Karlovich RS, Osier HA: Remote masking generated by high frequency tone complexes. Audiology 1977;16:507–521.

Antonio Quaranta, MD, Center of Audiology and Otology, ENT Department, University of Bari, I-70124 Bari (Italy)

Grandori F, Cianfrone G, Kemp DT (eds): Cochlear Mechanisms and
Otoacoustic Emissions. Adv Audiol. Basel, Karger, 1990, vol 7, pp 180–186

Intracochlear Mechanisms Involved in the Generation of Delayed Evoked Otoacoustic Emissions

Giovanni Rossi

Institute of Audiology, University of Turin, Italy

Every system that vibrates produces sounds, except in a vacuum. In 1948, Gold [1] declared that the inner ear, whose activity is based on a process of this type, could be a source of sounds. The instruments then available were incapable of discerning this phenomenon, however, and another 30 years passed before it was recorded for the first time by Kemp [2].

The production of delayed evoked otoacoustic emissions (DEOAEs) is attributed to the contractile activity of outer hair cells (OHCs) [3, 4]. This activity, which was demonstrated for instance by Flock [5], Brownell [6] and Zenner [7], results in a modification of the maximum amplitude of the travelling wave of the basilar membrane (BM).

The effects of OHC contraction would thus provide an explanation for two phenomena at the same time: DEOAEs and Evan's 'second filter' [8], regarded as the source of the particular frequency selectivity of Corti's organ (fig. 1). This view, however, fails to take into account the possibility that the active intracochlear mechanism could be accompanied by a passive mechanism attributable to the travelling wave of the BM following displacement of the perilymph by the stapes. If DEOAEs are a biological phenomenon caused by a vibratory process, they must also be present in subjects without Corti's organs, but with at least an unimpaired membranous part of the cochlear duct, i.e. the BM and Reissner's membrane.

The ideal situation in which to investigate the possible contribution of a passive intracochlear mechanism in the production of DEOAEs in man occurs in post-mumps and post-measles hearing loss. In these cases Corti's

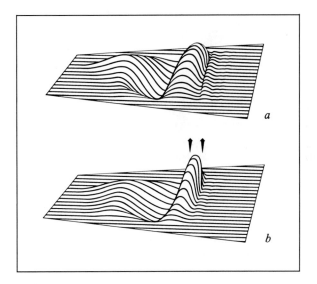

Fig. 1. The amplitude of the maximum oscillation of the travelling wave induced in the BM (*a*) by displacement of the perilymph is altered by the contractile activity of the OHCs (*b*).

organ has been completely destroyed, whereas the BM and Reissner's membrane are unimpaired [9, 10] (fig. 2).

In addition, if subjects with anacusis induced in this way are examined, a possible interference from an active intracochlear mechanism stemming from sectors of Corti's organ not destroyed by the virus can be ruled out.

This study was conducted with tone bursts (0.5, 1, 2 kHz; 3 ms: rise and fall time 1 ms, 31 stimuli/s, 2,048 acquisitions). The signal generator and the recording and analysis apparatus consisted of an Amplaid MK VI system. Further technical details can be found in Rossi et al. [11, 12].

Attention was first directed to 6 subjects with receptive hearing loss of unknown etiology. It was found that DEOAEs could be recorded with tone bursts even when the tonal threshold was > 50 dB HL for individual frequencies.

In agreement with Tanaka [13] and Tanaka et al. [14], this finding shows that DEOAEs can be evoked with *tone bursts* even for frequencies with a threshold greater than 30–40 dB HL [12], as opposed to the view of Bonfils et al. [15] for DEOAEs evoked by *clicks* in subjects with mean audiometric thresholds greater than those previously indicated.

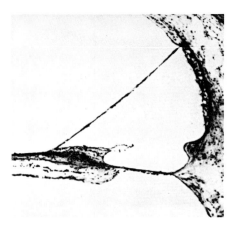

Fig. 2. Destruction of Corti's organ and preservation of the BM and Reissner's membrane in a case of post-mumps anacusis [from ref. 10].

The existence of a passive intracochlear mechanism in the production of DEOAEs was then investigated in 11 subjects aged 12–26 years with unilateral normal hearing and contralateral total hearing loss following mumps (8 cases) or measles (3 cases), who met a full 13 particular clinical and audiological requisites [12].

DEOAEs were recorded in all cases: for the three frequencies in 4, for the 0.5- and 1-kHz stimuli in 3, and for the 1- and 2-kHz stimuli in 4 (fig. 3). Given the same stimulus intensity above their threshold, the mean amplitude of the peak-to-peak DEOAEs on the deaf side was an average less than that on the hearing side [12] (fig. 4).

These results thus seem to suggest that DEOAEs are produced by the inner ear even in the absence of Corti's organ and thus necessarily also owe their origin to an intracochlear mechanism, namely BM oscillations passively induced by shifting of the perilymph by the stapes. It therefore lends support to part of the hypothesis underlying this research.

To check whether an active intracochlear mechanism could also be involved in the production of DEOAEs, a study was carried out in 20 subjects aged 19–23 with normal hearing. In this research, the relations between the behavior of the temporary threshold shift of 1-kHz tone bursts after 10 min of exposure to 90 dB HL with a 0.750-kHz tone, and the behavior of the DEOAE TTS under these conditions, have been investigated.

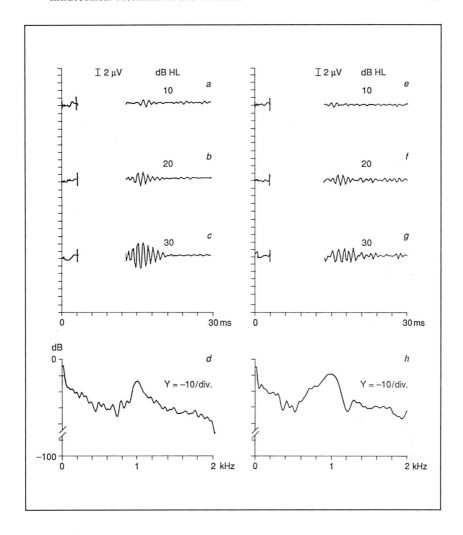

Fig. 3. Subject No. 3 (age: 16 years). Left post-mumps anacusis. Air conduction stimulation: 1-kHz tone bursts. 1-kHz DEOAE threshold of the normal ear: 8 dB HL; 1-kHz DEOAE threshold of the deaf ear: 22 dB HL. *a–c* DEOAEs from the normal ear. *d* Spectral analysis (FFT) of the DEOAEs in *c. e–g* DEOAEs from the deaf ear. *h* Spectral analysis (FFT) of the DEOAEs in *g*. The intensity of the stimulus is indicated as a value above the DEOAE threshold. The short vertical bar on the left indicates the onset of the acoustic stimulus whose registration is not shown [from ref. 12].

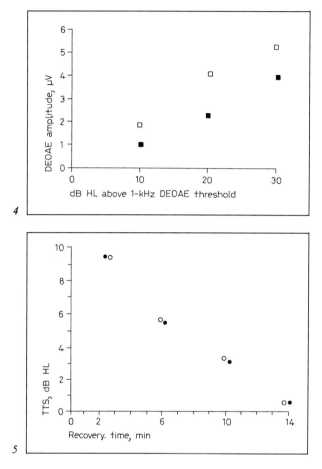

Fig. 4. Mean peak-to-peak amplitude of DEOAEs from 11 subjects with unilateral normal hearing (□) and controlateral post-mumps or post-measles anacusis (■).

Fig. 5. General and parallel pattern of the recovery time of the TTS of tone bursts and DEOAEs under the same experimental conditions [from Rossi et al., unpubl. data].

In both cases, the attenuator of the signal generator was calibrated in 3 dB HL steps and the DEOAE TTS was determined from one of the three tests that had to be performed within 2 min owing to reasons associated with the stimulation technique. The general pattern and the distinctly parallel relation between the two phenomena are illustrated in figure 5.

The DEOAEs are the expression of a phenomenon reaching the outer ear without involvement of the cochlear nerve fibers and synapses. DEOAE TTS can only be linked to the fatigue-induced alteration of hair cell functional activity. In DEOAE production, the contractile activity of OHCs may therefore constitute an active intracochlear mechanism which integrates the effects of a passive mechanism.

On the contrary, the tone burst TTS involves cochlear nerve fibers and synapses. The general and parallel pattern of the recovery time of both TTSs demonstrates that the fatigue-induced alteration of the hair cell functional activity certainly takes part in producing the tone burst TTS.

Acknowledgement

This research was supported by a grant from the Italian Ministry of Public Education (Project on Otoacoustic Emissions). The Editorial Offices of *Archives of Otolaryngology* and *Scandinavian Audiology* are gratefully acknowledged for permission to reproduce some figures from Smith and Gussen [10] and Rossi et al. [12].

References

1   Gold T: Hearing. II. The physical basis of the action of the cochlea. Proc R Soc Lond [Biol] 1948;135:493–498.

2   Kemp DT: Stimulated acoustic emissions from the human auditory system. J Acoust Soc Am 1978;64:1386–1391.

3   Bonfils F, Uziel A, Pujol R: Les otoémissions provoquées: une nouvelle technique d'exploration fonctionnelle de la cochlée. Ann Otolaryngol Chir Cervicofac 1987; 104:353–360.

4   Collet L, Disant F, Morgon A: Place des oto-émissions acoustiques dans l'exploration fonctionnelle auditive. Ann Otolaryngol Chir Cervicofac 1988;105:642.

5   Flock A: Contractile proteins in hair cells. Hear Res 1980;2:411–412.

6   Brownell WE: Observations on a motile response in isolated outer hair cells; in Webster WR, Aitkin LM (eds): Mechanisms of Hearing. Clayton, Monash University Press, 1983, pp 5–10.

7   Zenner HP: Motile responses in outer hair cells. Hear Res 1986;22:83–90.

8   Evans EF: Narrow tuning of the responses of cochlear nerve fibre emanating from the exposed basilar membrane. J Physiol (Lond) 1970;208:75P–76P.

9   Lindsay JR: Histopathology of deafness due to postnatal viral disease. Arch Otolaryngol 1973;98:258–264.

10 Smith GA, Gussen R: Inner ear pathologic features following mumps infection: Report of a case in an adult. Arch Otolaryngol 1976;102:108–111.

11 Rossi G, Solero P, Rolando M, Olina M: Delayed oto-acoustic emissions evoked by bone-conduction stimulation: experimental data on their origin, characteristics and transfer to the external ear in man. Scand Audiol 1988 (suppl 29): 1–24.

Alternative theories (called here spectro-temporal theories) propose that the pitch of complex tones is derived from both spectral and temporal information [5–8]; the spectral analysis performed in the auditory filters is followed by an analysis of the time pattern of the output of each filter, as represented in the patterns of phase locking in the auditory nerve. According to these theories, the pitches of complex tones can be extracted from unresolved as well as resolved components. Hence pitch perception and discrimination will not necessarily be impaired by removing the lower harmonics. Also, for high harmonics, which are not resolved by the auditory filters, the relative phases of the components may affect pitch perception [9]. For subjects with cochlear impairments, pitch perception may be impaired both because the temporal information conveying the pitch is more ambiguous [10] and because temporal analysis is impaired. The spectro-temporal theories also predict that pitch perception in impaired subjects should be more affected by relative phase than in normally hearing subjects, since the reduced frequency selectivity of the former means that even the low harmonics may interact at the output of the auditory filters [10].

The second experiment reported here tests the predictions of the spectral and spectro-temporal theories by measuring thresholds for the pitch discrimination of complex tones with and without the lower harmonics, for two different phase relationships of the harmonics.

In the third experiment, thresholds were measured for detecting a change in phase of a single component within a complex tone containing 20 equal-amplitude harmonics. All of the harmonics except one were added starting in cosine phase. The remaining harmonic started in cosine phase, but was shifted in phase half-way through either the first or the second of the two observation intervals in a trial. The task of the subject was to identify the interval in which the phase shift occurred. For higher harmonic numbers, provided the phase shift is sufficiently large, the phase-shifted component appears to 'pop out' and is heard as a pure tone against the background of the buzz-like cosine-phase complex [11, 12].

The most obvious explanation of this effect is in terms of the temporal waveform at the output of an auditory filter centred at the frequency of the phase-shifted harmonic. When all the components are in cosine phase, this waveform shows a distinct modulation, with low-amplitude portions between large peaks. This is illustrated in figure 1. When a single harmonic is phase shifted, that component appears in the low-amplitude portion between major peaks, as also shown in figure 1. Subjects may be able to detect the signal in these low-amplitude portions.

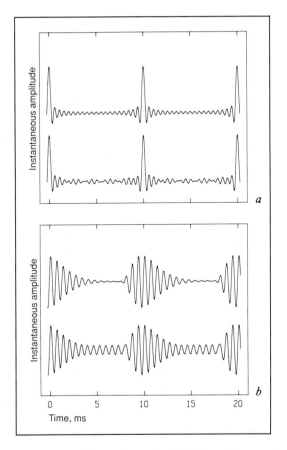

Fig. 1. a Waveforms of two stimuli (unfiltered): a complex tone containing the first 20 harmonics of a 100-Hz fundamental, all at equal amplitude and starting in cosine phase (top trace) and that same stimulus with the phase of the 15th harmonic advanced by 30° (bottom trace). *b* Output of a simulated auditory filter centred at 1.5 kHz in response to the two waveforms in the upper panel. For the cosine-phase stimulus the filter output shows a distinct low-amplitude portion. When the 15th harmonic is phase-shifted, an oscillation at the frequency of the harmonic appears in this low-amplitude portion.

This explanation suggests that sensitivity to a phase shift of a single component should depend both upon the frequency selectivity and the temporal resolution of the ear. A strong modulation of the waveform at the output of the auditory filter will only occur when several components fall within the passband of the filter; the more closely spaced the components,

Table 1. For each subject the table shows age, sex, absolute thresholds in the normal and impaired ears in dB HL at six frequencies (kHz), and the clinical diagnosis

Subject	Age	Sex	Ear	Threshold, dB HL						Diagnosis
				0.25	0.5	1.0	2.0	4.0	8.0	
FR	71	M	normal	20	15	5	10	20	35	
			impaired	40	45	60	55	45	65	cochlear loss
GB	59	F	normal	10	15	25	35	45	55	
			impaired	50	55	55	55	60	70	cochlear loss
PM	69	F	normal	20	15	10	15	30	60	
			impaired	65	60	55	50	55	70	Menière's
DJ	48	M	normal	5	10	0	5	15	10	
			impaired	65	70	60	55	70	80	Menière's
TD	72	M	normal	25	20	15	10	50	70	
			impaired	60	55	45	35	50	85	Menière's

relative to the auditory filter bandwidth, the more modulation will there be. Thus we might expect that normally hearing subjects would perform poorly for low harmonic numbers, whereas hearing-impaired subjects might be able to detect the phase shift even for low harmonic numbers. However, temporal resolution and absolute sensitivity are also involved, since the phase-shifted component has to be detected in the brief low-amplitude segments of the waveform at the output of the auditory filter. Impaired temporal resolution and reduced absolute sensitivity might offset the advantage for phase sensitivity produced by reduced frequency selectivity. Our third experiment tested these ideas.

General Method

Procedure

All thresholds were measured using an adaptive, two-alternative forced-choice procedure that tracks the 71% point on the psychometric function. All thresholds reported are based on the mean of estimates from at least three runs. Subjects were tested in a double-walled sound attenuating chamber. Stimuli were delivered using Sennheiser HD 414 earphones. When the impaired ear was being tested, a continuous pink noise with a spectrum level at 1 kHz of 25 dB was delivered to the normal ear.

Table 2. Equivalent rectangular bandwidths (ERBs) of the auditory filters for the normal and impaired ear of each subject, expressed as a proportion of centre frequency, for centre frequencies of 500, 1,000 and 2,000 Hz

Centre frequency	Ear	Subject				
		FR	GB	PM	DJ	TD
500 Hz	normal	0.20	0.22	0.17	0.21	0.18
	impaired	0.28	> 1.0	> 1.0	0.51	> 1.0
1,000 Hz	normal	0.14	0.19	0.15	0.15	0.14
	impaired	0.37	0.46	0.60	> 1.0	0.31
2,000 Hz	normal	0.15	0.32	0.21	0.17	0.18
	impaired	0.53	0.65	0.33	0.38	0.30

Subjects

Five subjects with unilateral cochlear hearing loss were tested in all three experiments. They were carefully screened to exclude conductive or retrocochlear losses. The impaired ears were selected to have moderate, relatively flat losses as a function of frequency. The normal ears had absolute thresholds close to normal limits given the ages of the subjects. Table 1 gives audiometric data for each subject. All subjects had extensive practice in psychoacoustic tasks.

Experiment 1: Measurement of Auditory Filter Shapes

Auditory filter shapes were estimated by measuring the threshold for a sinusoidal signal presented in a noise masker with a spectral notch. The noise spectrum level was 50 dB. The notch was placed both symmetrically and asymmetrically about the signal frequency, to allow the measurement of the asymmetry of the auditory filter. The method was exactly as described in [13] and the reader is referred there for details. We present here only one summary statistic of the derived auditory filters, namely their equivalent rectangular bandwidths (ERBs), expressed as a proportion of the centre frequency. These are presented in table 2.

The ERBs for the 'normal' ears were generally within the range expected for normal subjects [14], except for subject GB at 2,000 Hz. For each subject, the ERB for the impaired ear is greater than that for the normal ear at each centre frequency. For subjects GB, PM and TD at 500

Hz, and DJ at 1,000 Hz, there was too little frequency selectivity for the ERB to be estimated. Overall, the results confirm that the cochlear hearing losses of these subjects were associated with reduced frequency selectivity.

Experiment 2: Pitch Discrimination of Complex Tones as a Function of Harmonic Content and the Relative Phases of the Components

Stimuli

In this experiment we measured the thresholds for detecting changes in the fundamental frequency (F_0) of complex tones. Details of the method of stimulus generation and of the experimental procedure are given in Moore and Glasberg [15] and [16]. The complex tones had a mean F_0 of 200 Hz. They contained either harmonics 1–12 or 6–12, and the level of each harmonic was 70 dB SPL. For the complex tones with harmonics 6–12, thresholds were measured both in the presence and absence of a noise designed to mask combination tones in the region below the lowest harmonic. The noise was low-pass filtered at 800 Hz (96 dB/oct slope) and had a spectrum level of 35 dB.

The components were added in two different phase relationships. For the cosine-phase condition, all components started in cosine phase (90°), giving a very 'peaky' waveform. For the alternating-phase condition, the odd-numbered harmonics started in cosine phase and the even-numbered harmonics in sine phase (0°), giving a less 'peaky' waveform, but with two major peaks per period. Each tone had a steady-state portion of 200 ms and 10-ms rise/fall times, each shaped with the appropriate half-cycle of a raised-cosine function. The silent interval between the two tone bursts in a trial was 500 ms.

According to the spectro-temporal theories, the temporal structure of the waveforms at the outputs of the auditory filters is important for pitch perception. Simulations of auditory filtering showed that, for a filter centred at the fifth harmonic, the cosine-phase wave usually has well-defined envelope peaks which would be expected to give clear information about F_0. The alternating-phase wave, on the other hand, has a much flatter envelope, so that F_0 is less well represented. These differences between the waves are also found when the simulated filter is centred on higher harmonics. In the case of a filter with reduced frequency selectivity, as would occur in an impaired ear, the temporal structure at the output of the filter is very different for the alternating-phase wave and the cosine-phase wave, the alternating-phase wave having peaks separated by times not corresponding to the period of F_0. The spectro-temporal theories predict that such a wave should have an ambiguous pitch and should be poorly discriminated. It should be noted that if the phase responses of the auditory filters are markedly nonlinear, the waveforms after filtering would be quite different from those described above. However, we would still expect the different phase conditions to give rise to different waveforms.

Results

The results are presented in figure 2. Each panel compares results for a particular pair of conditions. Thresholds for discriminating a change in F_0 are expressed as a percentage of F_0. It turned out that the low-pass noise had very little effect on the thresholds for the complex with harmonics 6–12, and the results shown are averaged for conditions with and without the noise. For the normal ears, thresholds are typically about 0.5% or less, except for subject TD whose performance is slightly poorer. Thresholds are always higher for the impaired than for the normal ear for each subject and each type of complex tone. This is consistent with both the spectral and spectro-temporal theories of pitch.

Consider now the effect of harmonic content (fig. 2c, d). For the normal ears, the presence or absence of the lower harmonics has no consistent effect on the thresholds. This indicates that the lower harmonics are not essential for pitch discrimination. For the impaired ears, performance is better when the lower harmonics are *absent*. For every subject and for both phase conditions, thresholds are lower for harmonics 6–12 than for harmonics 1–12. The average ratio of thresholds is 2.2. This indicates that pitch information is conveyed most effectively by the higher harmonics, and that the lower harmonics somehow interfere with this information. Tests conducted with other subjects and with other fundamental frequencies have shown that this effect does not always occur [15, 16]. However, it does appear to be consistent for the subjects of this experiment. This is not easily explained by spectral theories of pitch perception. It is noteworthy that the four subjects (GB, PM, DJ and TD) who performed very poorly for the complex tones with harmonics 1–12 all had very poor frequency selectivity in the frequency region of the lower harmonics (500 Hz).

Consider now the effects of the relative phases of the components (fig. 2a, b). For the normal ears there are no consistent effects of relative phase. For the impaired ears, there is a consistent effect for the complex

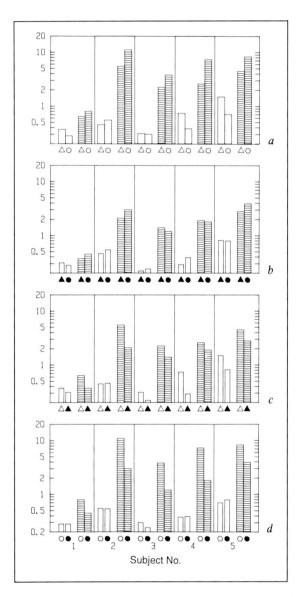

Fig. 2. Results of experiment 2, showing thresholds (DLs) for the discrimination of F_0, plotted as a percentage of F_0 on a logarithmic scale. Each panel (*a–d*) compares a particular pair of conditions. Results for the normal ears are shown by open bars, and those for the impaired ears by shaded bars. The four conditions were as follows: harmonics 1–12 cosine phase (△); harmonics 1–12, alternating phase (○); harmonics 6–12, cosine phase (▲); harmonics 6–12, alternating phase (●).

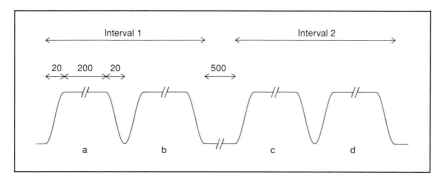

Fig. 3. Illustration of the structure of a single trial in the two alternative forced-choice task of experiment 3. a and c are always in cosine phase. Either b or d has a phase-shifted component.

tones with harmonics 1–12 (fig. 2a). The alternating-phase condition gives higher thresholds than the cosine-phase condition: the average ratio of thresholds is 1.85. This effect is in the direction predicted by the spectro-temporal theories. For the complex tones with harmonics 6–12, subjects FR, GB and TD show an effect in the same direction, but subjects PM and DJ show no effect of phase. Overall, the results are consistent with the prediction of the spectro-temporal theory that changes in the relative phases of the components should have greater effects on pitch discrimination for impaired ears than for normal ears.

Experiment 3: Difference Limens for Phase

Stimuli

Details of the method of generating the stimuli and of the procedure for measuring thresholds are given in [17]. The stimuli were complex tones containing the first 20 harmonics of 100 Hz. All harmonics had a level of 70 dB SPL. The temporal structure of a trial is illustrated in figure 3.

There were two observation intervals, and each interval was divided into two segments, giving four segments in total: a, b, c and d. Each segment had an envelope with a 200-ms steady state portion and 20-ms ramps, shaped according to a raised-cosine function. Segments a and b (interval 1) and segments c and d (interval 2) were adjacent in time. The interval between the two observation intervals (from the end of b to the start of c) was 500 ms. The first two segments in each interval (a and c) had harmonics which started in cosine phase. Either segment b or segment d contained a component which was advanced in phase. The remaining components in that segment, and all the components in the remaining segment started in cosine phase. The task of the subject was to identify the interval containing the phase-shifted component. The task was designed in this way so that the phase shift could be identified on the basis of any audible change. The subject did not have to label the stimuli in any way; all that was required was the detection of a change *within* an observation interval.

Results

The results are shown in figure 4. Each panel shows data for the normal ear and impaired ear of one subject. Figure 4f shows the (geometric) mean results. Note that if a phase shift could not be heard for shifts greater than 90°, then it could not be heard at all. Points plotted at 90° on the ordinate indicate that a threshold could not be determined, since the phase shift could not be heard. For the normal ears, thresholds were usually lowest for harmonic numbers between about 10 and 18. For the lowest harmonic numbers, the phase shift could not usually be heard at all. This is consistent with the idea that phase shifts for these stimuli are only heard when several harmonics interact within an auditory filter.

For the impaired ears, thresholds were generally much higher than for the normal ears. This is not surprising, since performance on this task worsens markedly as the sensation level (SL) of the stimuli decreases [17]. The components were at much higher SLs in the normal than in the impaired ears. To assess the effect of SL, phase thresholds for the 10th and 15th harmonics were measured for the normal ears of GB, PM and DJ, using stimuli whose components were at the same SL as would be produced by the 70 dB/component stimulus in the impaired ear. The phase changes could not be detected at these low sensation levels. Thus, at comparable SLs, the impaired ears showed *greater* phase sensitivity than the normal ears.

The normal and impaired ears also differed in the pattern of results across harmonic numbers. This is illustrated in table 3, which shows phase thresholds averaged separately (geometric means) for harmonics 10–18 and 5–8. For the normal ears, thresholds were always higher for the low harmonic numbers. For the impaired ears of FR and DJ the opposite was true; thresholds were lower for the low harmonic numbers. Indeed, for

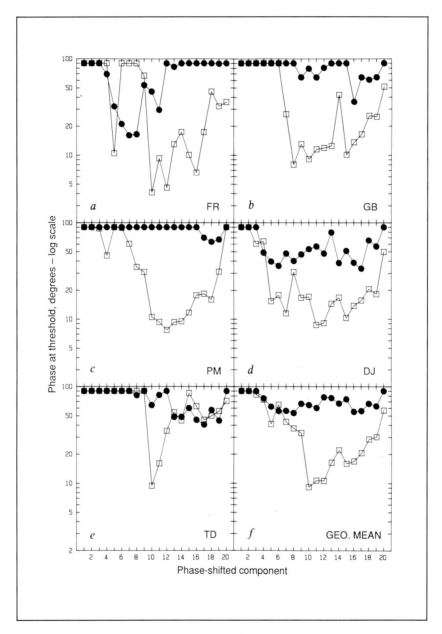

Fig. 4. Results of experiment 3 showing the phase shift required for threshold in degrees (log scale). □ = Normal ears; ● = impaired ears. *a–e* Results for individual subjects. *f* Geometric mean results.

Table 3. Thresholds (in degrees) for detecting shifts in phase of a single harmonic of a complex tone, averaged for harmonics 5–8 and 10–18

Harmonics averaged	Ear	Subject				
		FR	GB	PM	DJ	TD
5–8	normal	53+	36+	64+	18	CND
	impaired	21	CND	CND	31	CND
10–18	normal	11	15	12	14	38
	impaired	73+	70+	84+	50	60

CND stands for 'could not do' and indicates that phase changes could not be detected for any harmonics within the specified range. Some entries are averages including harmonic numbers for which phase shifts could not be detected. In those cases, indicated by a +, indeterminate thresholds were assigned a nominal value of 90°.

harmonic numbers 6–8, FR showed lower phase thresholds in his impaired than in his normal ear, in spite of the lower SL in his impaired ear. This is consistent with the idea that the broader auditory filters in the impaired ear can give rise to greater phase sensitivity than normal for low harmonic numbers. However, it is noteworthy that the two subjects showing good phase sensitivity in their impaired ears for low harmonic numbers (FR and DJ) actually had better frequency selectivity in the frequency region of those harmonics (500 Hz) than the other subjects.

General Discussion and Conclusions

Our subjects were carefully selected so as to have audiograms which were as similar as possible; the impaired ears all had moderate losses which were relatively flat across frequency. In spite of this, there were marked individual differences across subjects in all three experiments. This may partly reflect differences in the underlying pathology. PM, DJ and TD, who were diagnosed as having Menière's disorder, all had poorer frequency selectivity at 500 Hz than at 2,000 Hz. This was not true of FR who showed relatively good frequency selectivity at 500 Hz, but rather poor selectivity at 2,000 Hz. However, even within the three subjects with Menière's disorder there are marked individual differences.

Two aspects of the results of experiment 2 support spectro-temporal theories of pitch perception as opposed to spectral theories. Firstly, pitch discrimination was not impaired by removing the lower harmonics. Indeed, for the impaired ears, performance was improved by removing the lower harmonics. This suggests that pitch can be extracted using temporal information from the higher, unresolved harmonics. The reason why the lower harmonics had a deleterious effect is unclear. One possibility is that adding extra harmonics produced more complex waveforms at the outputs of the auditory filters, making the temporal information more ambiguous [10]. A second possibility is that the temporal information from neurones tuned to the lower harmonics somehow conflicted with the information from neurones tuned to the higher harmonics.

The phase effects found in experiment 2 also support the spectro-temporal theories. For the complex tones with harmonics 1–12, pitch discrimination in the impaired ears was worse when the harmonics were in alternating phase than when they were in cosine phase. This effect was relatively small for FR, who had reasonably good frequency selectivity at 500 Hz, but was large for the remaining subjects, who showed much poorer frequency selectivity at 500 Hz.

The results of experiment 3 showed that the ability to detect a change in phase of a single component in a complex tone was generally worse for the impaired than for the normal ears when the comparison was made at equal sound pressure levels. However, the impaired ears were better than normal when compared at equal SLs. For harmonic numbers 6–8, subject FR showed lower phase thresholds in his impaired than in his normal ear, in spite of the lower SL in the impaired ear. Two of the subjects, FR and DJ showed better phase sensitivity for low harmonic numbers than for high in their impaired ears, the opposite of the pattern found for the normal ears. These results confirm that reduced frequency selectivity can lead to enhanced phase sensitivity, although frequency selectivity is clearly not the only factor determining phase sensitivity.

It appears that good phase sensitivity is associated with poor, but not grossly impaired, frequency selectivity. For example, of the impaired ears, subject FR showed the greatest phase sensitivity at low harmonic numbers, and he had the best frequency selectivity at 500 Hz, the frequency region of the lowest harmonics. The three subjects with no measurable frequency selectivity at 500 Hz, GB, PM and TD, were all unable to detect phase changes in the lower harmonics.

broadening of frequency selectivity associated with tinnitus offers another objective way of assessing impairment induced by tinnitus.

The present article reports progress made in the continuation of our previous study. Attention was specially focused on patients presenting tinnitus unilaterally and rather equivalent audiograms bilaterally in order to minimize intersubject variability and to obtain more directly comparable results.

Materials and Methods

Among 26 subjects tested until now, only 12 presented tinnitus in one ear and a hearing loss approximately similar for both ears. Only these 12 subjects will be considered in this report. All tests were performed in a sound-proof booth; stimuli were generated by a computer and delivered to the ear through electrostatic headphones. In a first step, pitch and loudness matching of tinnitus were performed using pure tones presented ipsilaterally. Matching in most cases was to a high frequency and a level only about 10 dB above audiogram threshold similarly to previous studies [4]. Tinnitus pitch was matched to frequencies between 3 and 5 kHz for 5 subjects and to frequencies between 7 and 9 kHz for the other 7 subjects. In all 12 cases an audiogram was measured in third-octave steps at high frequencies.

To define frequency selectivity, a simultaneous masking paradigm was used. An objective test tone at the frequency match of tinnitus was presented at 5 dB above tinnitus loudness matching. The test tone was trapezoidally gated with 25 ms for rise and fall times and 50 ms of plateau. It was presented at a rate of 5/s during 5 s. It was always declared quite distinguishable from tinnitus by all patients. Masking pure tones were presented in third-octave intervals and in 5-dB steps. For several subjects the curve of pure-tone masking of tinnitus was also obtained. In addition, for control, a curve of frequency selectivity was taken from the ear with tinnitus at a frequency different from tinnitus.

Results

In all 12 cases with similar audiograms for both ears a hearing loss at high frequencies was present. An example of data obtained from one subject is presented in figure 1. A clear broadening of frequency selectivity can be seen to be associated with tinnitus. In this case, the broadening seems mostly due to a reduced slope on the low-frequency side. It must be noted that the selectivity was better for the 1,000-Hz test stimulus where no tinnitus was present. Similar broadenings of frequency selectivity were observed for 11 of the 12 subjects with equivalent audiograms bilaterally in agreement with our previous report.

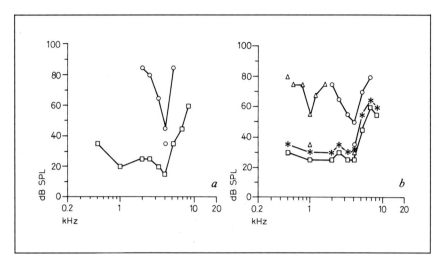

Fig. 1. Examples of data obtained in one patient. *a* Data from the nontinnitus ear, with a frequency selectivity curve (○) obtained for the same test tone as that matched to tinnitus in the other ear. *b* Ear with tinnitus. The tinnitus match (▲) was 4,120 Hz and 30 dB SPL. The pure-tone masking curve (*) of tinnitus showed no frequency selectivity and followed the audiogram (□). Two frequency selectivity curves were obtained from that ear: one for a test tone at tinnitus frequency (○) and the other for a 1,000-Hz tone (△).

Further inspection of the results revealed that the broadening was correlated with the hearing loss at tinnitus frequency: the higher the threshold elevation, the larger the difference in frequency selectivity between the two ears. To quantify this, we measured the ratio of the bandwidths expressed in octave at 10 dB above the tip of the frequency selectivity curve (BW10) for the tinnitus ear relative to the nontinnitus ear. In the example of figure 1, BW10 for the tinnitus ear was 0.78 octave and for the nontinnitus ear it was 0.31 octave. This corresponds to a broadening factor of 2.5 (0.78/0.31). Such broadening factors were calculated for the 12 subjects with approximately similar audiograms bilaterally and plotted as a function of hearing loss at the tinnitus frequency. The data are presented in figure 2. It can be clearly seen that the broadening factor increases with hearing loss. As is well known [9], and as reported in our previous study, BW10 for nontinnitus ears increases with hearing loss. The additional broadening associated with tinnitus is revealed in this study to be proportional to the hearing loss as well.

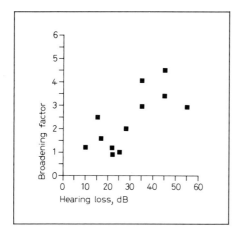

Fig. 2. Values of tinnitus-associated broadening factor as a function of hearing loss for the 12 subjects.

Discussion

In a previous study we have established that frequency selectivity for an objective tone at tinnitus frequency is broader than could be expected from hearing loss [8]. In the present article we showed that the tinnitus-associated perturbation of objective tone perception increases with the degree of hearing loss. These data indicate that the pathology underlying tinnitus also gives rise, in the frequency zone of tinnitus, to impairments in the perception of objective tones. The increase of this impairment with the increase of hearing loss suggests that both could have some common origin. Since this tinnitus-associated impairment is monaural the origin could be the cochlea. On the other hand, tinnitus-associated impairment further deteriorates selectivity increasingly with hearing loss. This suggests that the tinnitus-associated broadening might involve a process different from the broadening usually observed with hearing loss. It can be imagined that tinnitus deteriorates signal detection by adding background noise. It is indeed well established that the broadening of frequency selectivity is accompanied by a decrease in the detection of a signal in noise [10]. Studies of patients with sporadic tinnitus could help clarifying this issue. Further studies seem worth performing for instance to investigate the extent in frequency of this tinnitus associated broadening which might vary from one subject to another as does the efficient bandwidth of masking noise for tinnitus [4].

In a clinical perspective, and alteration of objective tone perception associated with tinnitus appears as a useful complementary phenomenon to measure when assessing therapies for tinnitus. In particular, it seems interesting to examine possible covariations with the subjective annoyance of tinnitus and its matching to an objective sound.

References

1 Evans EF, Borerwe TA: Ototoxic effects of salicylates on the responses of single cochlear nerve fibers and on cochlear potentials. Br J Audiol 1982;16:101–108.
2 Goodey RJ: Drugs in the treatment of tinnitus; in Tinnitus Ciba Symposium. London, Pitman Press, 1981.
3 Vernon JA, Meikle MB: Tinnitus masking: unresolved problems; in Tinnitus Ciba Symposium. London, Pitman Press, 1981.
4 Cazals Y, Bourdin M: Etude acoustique des acouphènes. Rev Laryngol 1983;104: 433–438.
5 Cazals Y, Rouanet JF, Negrevergne M, Lagourgue P: First results of chronic electrical stimulation with a round window electrode in totally deaf patients. Arch Otorhinolaryngol 1984;239:191–196.
6 Cazals Y, Negrevergne M, Aran JM: Electrical stimulation of the cochlea in man: hearing induction and tinnitus suppression. J Am Audiol Soc 1978;3:209–213.
7 Feldmann H: Homolateral and contralateral masking of tinnitus by noise bands and pure tones. Int Audiol 1971;10:138–144.
8 Dauman R, Cazals Y: Auditory frequency selectivity and tinnitus. Arch Otorhinolaryngol In press, 1989.
9 Florentine M, Buus S, Scharf B, Zwicker E: Frequency selectivity in normally hearing and hearing impaired listeners. J Speech Hear Res 1980;23:646–669.
10 Dreschler WA, Plomp R: Relation between psychophysical data and speech perception for hearing impaired subjects. I. J Acoust Soc Am 1980;68:1608–1615.

Yves Cazals, MD, Laboratoire d'Audiologie Expérimentale,
Inserm unité 229, Université de Bordeaux, Service ORL, Hôpital Pellegrin,
F–33076 Bordeaux (France)

Subject Index

Acetylcholine 171, 172
  receptors 39
Acetylsalicylic acid 149
Acoustic cochleography 78
Acoustic oscillation 86
Acoustic stimulation 42, 43
Actin 172
Advanced cochlear echo system 79
Alternating wave phase 193–195
Amacusis 181, 183, 184
Aminoglycerides, sensory hair cells 42–45
Amphibian papilla 50, 53
Amplitude 111–116, 120–122
  response 103
Atropine 171–176
Audiograms 202
Auditory filters 187–192

Basilar membrane 1–3, 6, 7, 9, 36, 39, 126
  otoacoustic emissions 47, 72, 117, 180
  vibrational response 13–15
Basilar papilla 50
Burst frequencies 101

Caffeine, cochlea mechanics 27, 31–34
Calcium, outer hair cell motility 38
Cilia 10
Click stimulation 43, 47, 81, 101–106, 150, 181
Click-evoked emissions 78, 91, 94
Cochlea 171
  distortion-evoked otoacoustic emissions 139–147

mechanical responses 13–24
micromechanics 27–34
sandwich modelling 2–11
Cochlear pattern, sensorineural impairment 151–153
Cochlear response 82, 83, 85
Cochlear waves 2–11
Continuous tonal stimulation 48, 50
Contralateral auditory stimulation 164–169
Contralateral efferent system 165
Contralateral sound 157–160, 164
Cosine phase 188, 192
Coupling resonance 87–89
Critical band rate 67
Critical ratio 173, 174, 176
Cubic difference tones 126–136
Cuticular plate 35, 37, 172

Deconvolutions 100–108
Delayed evoked otoacoustic emissions 66–68, 77, 180–185
Detection threshold 141, 142
Distortion product(s) 117–136, 156–161
  frequencies 52
  otoacoustic emissions 68, 70, 72, 117–124, 126, 128–136, 139–147
Distortion tone 57–62
Dummy coupler 68, 69

Ear canal
  otoacoustic probe 78, 83–87
  responses 80